Climate Change: The Ethics of Adaptation

Climate Change: The Ethics of Adaptation

Edited by Mary D'souza

STATES
ACADEMIC PRESS
www.statesacademicpress.com

States Academic Press,
109 South 5th Street,
Brooklyn, NY 11249, USA

Visit us on the World Wide Web at:
www.statesacademicpress.com

ISBN: 978-1-63989-688-2

Cataloging-in-publication Data

Climate change : the ethics of adaptation / edited by Mary D'souza.
p. cm.
Includes bibliographical references and index.
ISBN 978-1-63989-688-2
1. Climatic changes. 2. Climatic changes--Moral and ethical aspects. 3. Global warming--Moral and ethical aspects. 4. Environmental ethics. 5. Adaptation (Biology). I. D'souza, Mary.
QC903 .C55 2023
363.738 74--dc23

TABLE OF CONTENTS

Preface

Climate change refers to the shift in temperature over a prolonged period of time. The process of adjusting to the changes in climate is known as climate change adaptation. The actions taken to adapt to climate change can be categorized as either transformative or incremental. The former responds to climate change by trying to change the fundamental attributes of a system while the latter seeks to maintain the integrity of a system. One of the important factors which need to be considered while making policies related to climate change adaptation is adaptive capacity. It encompasses the ability of any system to adjust to climate change in order to reduce damage or deal with the outcomes. This book contains some path-breaking studies on climate change and the ethics of climate change adaptation. It will serve as a reference to a broad spectrum of readers.

This book is the end result of constructive efforts and intensive research done by experts in this field. The aim of this book is to enlighten the readers with recent information in this area of research. The information provided in this profound book would serve as a valuable reference to students and researchers in this field.

At the end, I would like to thank all the authors for devoting their precious time and providing their valuable contribution to this book. I would also like to express my gratitude to my fellow colleagues who encouraged me throughout the process.

Editor

1

Farm-Level Impacts of Greenhouse Gas Reductions for the Predominant Production Systems in Northern Nigeria

Taiwo B. Ayinde, Benjamin Ahmed and Charles F. Nicholson

Contents

Abstract

This chapter summarizes the sources of greenhouse gas (GHG) emissions from different economic sectors in Nigeria and emphasizes those arising from agriculture and forestry. The impacts of climate change on agricultural systems in

T. B. Ayinde (✉)
Samaru College of Agriculture, Division of Agricultural Colleges, Ahmadu Bello University (ABU), Zaria, Nigeria

B. Ahmed
Department of Agricultural Economics, ABU, Zaria, Nigeria

C. F. Nicholson
Nijmegen School of Management, Radboud University, Nijmegen, Netherlands

Nigeria are likely to be large, motivating the need for additional knowledge to assess current practices and formulate appropriate modifications for both mitigation and adaptation. Some current farming practices are believed to be adaptive, but further study would provide better assessments. We also analyzed the trade-offs between household income and GHG emissions at two contrasting sites in northern Nigeria. A farm optimization model maximizing the value of crop, livestock, and tree production activities in a single representative year assessed the potential impacts for GHG reductions of 10% and 25% and the maximum allowable reductions of 26% and 30% on farm activities and income. Emissions reductions of 10% reduced annual household incomes by less than 5% but required substantive changes, especially in livestock owned. Maximum possible GHG emissions reductions (while still meeting minimum household consumption needs) would require marked changes in production pattern and would lower household incomes by 22–44%. We did not assess effects over longer periods, where the role of livestock as a key asset may imply additional negative impacts. Productivity-enhancing technologies that would simultaneously reduce GHG emissions and increase incomes are needed for smallholder farms to play a larger role in climate change mitigation without the burden of reduced incomes and greater risk. This suggests the need for programmatic and policy actions both by national agricultural research systems and the Consultative Group for International Agricultural Research (CGIAR).

Keywords

Climate change · Agriculture · Nigeria

Introduction

Agriculture is the major employer of labor in Nigeria (Hansen et al. 2017) for over 70% of the estimated 200 million inhabitants (United Nations Department of Economic and Social Affairs: Population Division 2020). Almost 80% of over 70 million hectares of Nigerian land areas are under rainfed agricultural production that translates to about 40% of national Gross Domestic Product (GDP) (World Bank 2020). Agriculture is also a major contributor to greenhouse gas (GHG) emissions, as described below.

Overview of Greenhouse Gas Emissions from Agriculture in Nigeria

Agriculture was the largest of the four major sectors that contributed to Nigeria's GHG emissions in 2015. Agriculture, Forestry, and Other Land Use (AFOLU) accounted for two-thirds of total net national emissions of Teratonnes (Tt) CO_2 equivalent (CO_2-eq). The AFOLU sector includes emissions from land use, livestock, and removals from harvested wood products. Land use accounted for an estimated 421 Tt CO_2-eq, or about 90% of emissions from AFOLU. One principal source of land-use emissions is methane (CH_4) and nitrous oxide (N_2O) from burning of rice, maize, sugarcane, and wheat residue biomass. Other sources include direct and indirect emissions of N_2O from nitrogen-based synthetic fertilizers used in managed soils, indirect emissions of N_2O from manure management, and direct CH_4 emissions from rice cultivation. For livestock, enteric fermentation and manure management for cattle, sheep, and goats contributed 29 Tt CO_2-eq, with the remaining amounts from camels, mules, swine, and horses. Enteric fermentation was 90% of total livestock-related emissions, with the remainder from paddock, range, and pasture and solid storage systems of manure management. N_2O and methane CH_4 were the most important GHG from manure management.

Total GHG emissions continue to increase. In 2019, the United States Aid for International Development (USAID) reported a 25% increase of 98 Mt. CO_2-eq in the country's overall GHG emissions from 1990 to 2014. The average annual change in total emissions was 1% and was mainly from the land-use change and forestry (LUCF) sector. Similarly, a trend assessment from the period 2000 to 2015 showed that AFOLU remained the highest emitter of GHG throughout the years under review (Federal Republic of Nigeria 2018). Emission from land-use category, forest land remaining forest land remained the principal source, and it increased by 25% during the period, whereas removals through HWPs declined by 28%. There was an estimated 37% increase in emissions from rice cultivation, livestock production, manure management, and nitrogen-based synthetic fertilizers in managed soils over the entire period. The increase was attributed to the increase in livestock populations and more utilization of nitrogen-based synthetic fertilizers in managed soils as well as forest biomass loss. Thus, Nigeria's emissions are projected to grow to around 900 million tons per year in 2030, which translates to around 3.4 tons per person (Change Climate Response Policy 2015).

Impacts of Climate Change on Agriculture in Nigeria

Changes in climate, in particular, the frequency or severity of extreme events, are likely to have serious implications for rainfed agricultural production. Given the diversified socioeconomic conditions of agriculture in Nigeria, climate change could affect crop yields, disease patterns, and compound ecological disasters that include quelea bird's invasion, floods, and drought (Ajetomobi et al. 2015). However, the impacts are likely to vary by region. A vulnerability analysis conducted in 2014 by Nigerian's Federal Ministry of Environment indicated that states in the north experience higher degrees of vulnerability to climate change than those in the south (Haider 2019).

Many crops are sensitive to the modest changes in rainfall and temperature (Haider 2019). For example, the annual mean temperature needed for photosynthesis and growth is about 29 °C for maize and soybean. Currently, mean annual temperatures ranges from 30 °C to 34 °C. A mean annual temperature of 33 °C is desirable for cotton, and temperature above this threshold can be harmful to plant development (Ajetomobi et al. 2015). The combination of increasing temperature and lower rainfall has hastened desert encroachment, with loss of the wetlands, and rapid reductions in the amount of surface water, flora, and fauna resources on land. Kebbi and Jigawa states observed 5–10% reductions in millet yields due to Quelea bird invasion, in addition to drought in Katsina, whereas Sokoto, Zamfara, Kaduna, and Kano states reported between 10% and 25% loss in groundnut yield from heavy floods (National Agricultural Extension and Research Liaison Services (NAERLS) & Planning Research and Statistic Department (PRSD), 2012). Rainfall intensity and distribution have been altered from the usual 500–1000 mm per year. Flooding in 2012 resulted in 363 deaths, massive economic loss, and the displacement of more than 3.8 million people. The total value of destroyed physical and durable assets was reported to be 1.48 trillion (US$9.5 billion) or about 2% of the rebased GDP of US$510 billion (Change Climate Response Policy 2015).

Under the business-as-usual scenario of the Change Climate Response Policy (2015), Nigerian agricultural productivity could decline by up to 50% by 2080, with reductions in GDP as large as 4.5% even by 2050. Consequently, in the absence of mitigating measures, there would be steady depletion of vegetation and grazing resources in affected regions. This could also prompt massive emigration and resettlement of people to areas less threatened by desertification. Subsequently, exacerbating communal clashes among herdsmen and farmers and inter-ethnic clashes (Haider 2019).

Adaptation and Mitigation Strategies for Nigerian Agriculture

Climate change mitigation is the improvement of agricultural production practices to reduce GHG emission, while adaptation involves the uptake of agricultural practices to be more suitable for a modified climate in a particular location (IPCC 2007). Nigeria is now implementing the reduction of emissions due to deforestation and forest degradation (REDD) (USAID 2019), through the national community-based forest resources management program. Other mitigations and adaptation strategies include the adoption of climate-adapted crops (e.g., drought-tolerant and early maturing varieties of crops) and in the livestock sector, improved feed, pasture, ranch, and paddock management systems (Change Climate Response Policy 2015). Programs are also being implemented to reduce the volume of irrigation water, while making more effective use of rainwater and groundwater. Other efforts are directed at providing early-warning seasonal climate forecasts to facilitate adaptation in a given growing season.

However, such adaptation and mitigation strategies are location- and context-specific and are described as autonomous or opportunist adaptation (Haider 2019). Planned adaptation strategies require additional knowledge about appropriate levels of inputs, tillage practices, and assessment of the future environmental conditions (e.g., rainfall, temperature, and relative humidity). Research to assess and implement actions for both adaptation and mitigation. Simulation modeling can support assessment of adaptation strategies, particularly when experimental data are limiting and future environmental conditions uncertain (Nicholson et al. 2011; Kopainsky and Nicholson 2015). Consistent with the ongoing need for site-specific quantitative studies, the remainder of this chapter describes the use of modeling to assess the impacts of potential mitigation strategies on smallholder farming systems in the Northwestern area of Nigeria (Ayinde 2019).

Farm-Level Impacts of GHG Emissions Mitigation Strategies in Sudano-Sahelian and Sudan Savanna Ecological Zones in Northwestern Nigeria

An important question concerns the trade-offs among agricultural production, household incomes, and environmental impacts. To date, there are few studies that examine what actions would be required to reduce GHG emissions from smallholder agriculture production systems, and to understand the implications for household well-being. Northwestern Nigeria provides a conducive context for quantitative assessment of farm-level GHG mitigation strategies because it is a highly degraded region with higher degrees of vulnerability to climate change (Haider 2019). Within this region, differences in agro-ecology and production systems necessitate separate assessment of GHG-reduction strategies. Ayinde (2019) thus explored the production practices at two sites, the Bunkure (Sudan Savanna ecology) and Maigateri (Sudano-Sahelian ecology) local government areas (LGAs) located in Kano and Jigawa States, respectively (Fig. 1). The objective of the analysis was to describe predominant farming systems and discuss the extent to which they are consistent with either adaptation or mitigation strategies, and to evaluate quantitatively the impacts on agricultural production and potential trade-offs between household revenue and GHG emissions.

Kano and Jigwa differ in multiple respects, which facilitates a comparative analysis of mitigation strategies at the farm level. Kano is the most-extensively irrigated state in the country for the cultivation of rice, sugarcane, and vegetables, which are important emitters of GHG (IPCC 2013; Dunkelberg et al. 2014). Jigawa is characterized by shorter rainfall duration that does not support irrigated farming but sustains hardier shrubs and trees. In the LGAs of both states, however, mixed farming of grain and legume crops, trees, and livestock keeping are practiced.

Ayinde (2019) used participatory rural appraisal (PRA) and key informant interviews to characterize existing production systems at both study sites. Data were collected on the economic and environmental performance of the predominant production systems, land preparation, retaining of trees on crop field, use of trees branches and stem for firewood, use of crop residue as mulch, modified planting dates, use of early maturing crop variety, and agrosilvopastoral system. Emphasis were on trees that could serve multiple purposes of dry season feed, enriching soil Nitrogen, increasing protein and energy content in fodder while mitigating GHG emissions. This information was necessary to assess whether the production practices employed aligned with either a mitigation or adaptation strategy. The key

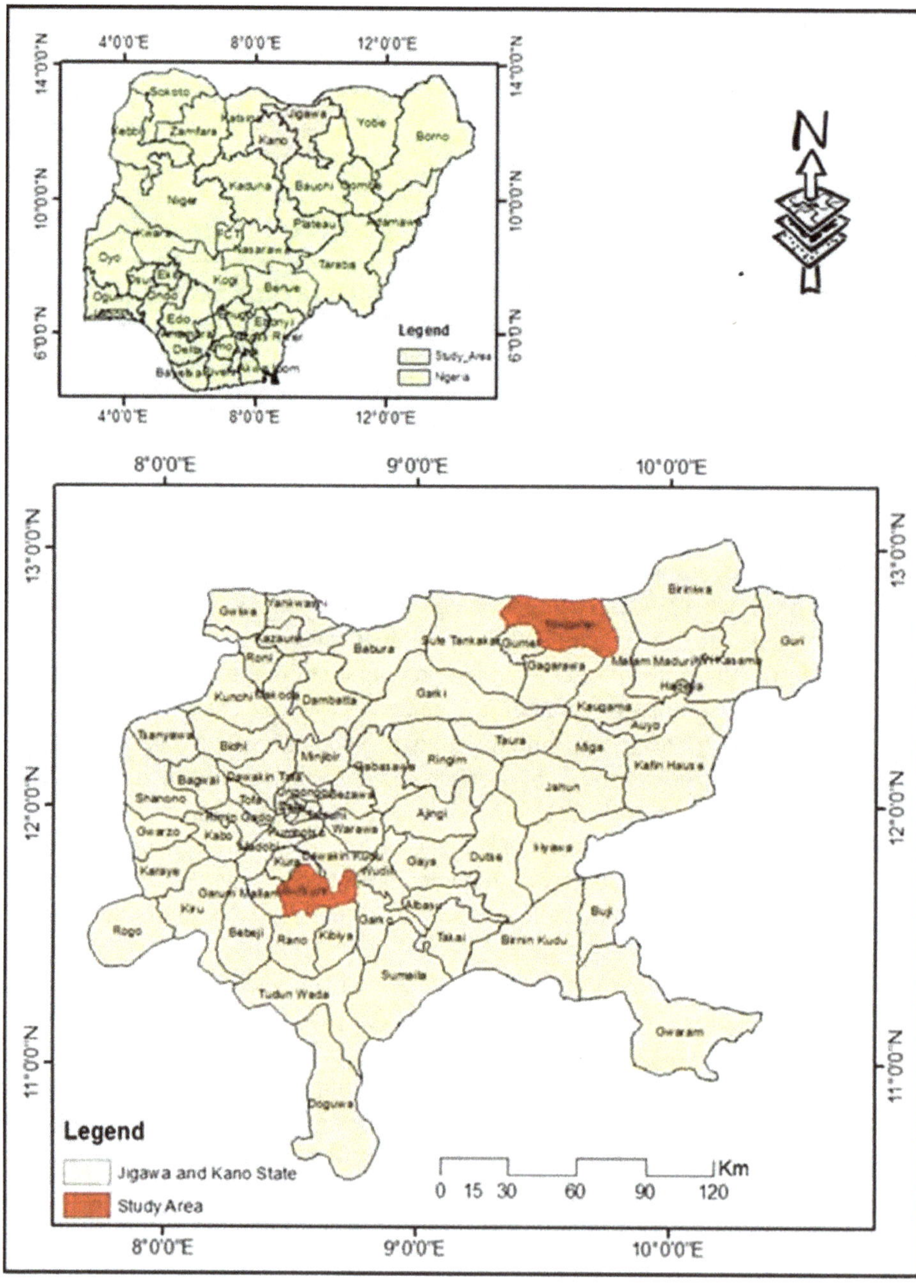

Fig. 1 Locations of the study sites Maigateri in Jigawa State and Bunkure in Kano State

informant interviews provided data of estimates of household-level expenditures on tree, crop, and livestock activities (See the assumed structure of inputs and outputs in the tree-crop-livestock production optimization model Appendix Table A1), the cost of animal draught, and the cost of manure applied. The description of the key specific inputs used in the current production practices were necessary to estimate the GHGs emissions (CH_4, N_2O, CO_2, and non-CO_2) in Appendix Tables A2.1 and A2.2.

The emission factors (EF) used to estimate GHG emissions to specific farm activities were obtained from IPCC Tier 1 default equations as well as methods from Standard Assessment of Mitigation Potentials and Livelihoods in Smallholder System (SAMPLES; www.samples.ccafs.cgiar.org) that corresponds to the predominant production systems in the study region. Data on dry matter (DM), metabolizable energy (ME), and crude protein (CP) that corresponded to the tree and crop yield levels were obtained from secondary sources such as feed composition tables or journal articles and grey literature including Feedipedia; FAO source: (http://www.fao.org/ag/againfo/themes/documents/PUB6/P620.htm); and (Dupriez and De Leener 1998).

To assess the potential impacts on agricultural production patterns and full household income scenarios evaluated a baseline (current practices and emission levels) and required GHG reductions of 10%, 25%, and the maximum reductions consistent with meeting assumed household consumption needs (26% and 30% reductions compared to the baseline). See the detailed description of the optimization model used to assess trade-offs between household welfare (income) and farm-level GHG emissions in Appendix 1.

Current Production Practices as Adaptation or Mitigation Strategies

A key finding was that farmers were aware of the multiple uses and economic value of trees for wood, food, fuel, fodder, soil protection, and soil reclamation (Table 1),

Table 1 Tree and crop (cereal and legume) production practices reported by households in Bunkure and Maigateri

	Bunkure		Maigateri	
	N = 50		N = 55	
Crop type and production practice	Frequency	Percentage	Frequency	Percentage
Trees				
Retained trees on crop field	50	100	55	100
Firewood	19	38	11	20
Cereal				
Predominant cereal grown	26	52	29	53
Use of improved sorghum varieties	11	22	16	29
Construction	13	26	14	25
Fuel	11	22	11	20
Legume				
Predominant legume grown	30	60	29	53
Sell	45	90	42	76
Fodder to own livestock	4	8	10	18
Soil amendment	1	2	3	6

similar to previous studies (Change Climate Response Policy 2015; UN-REDD Programme 2015). Consequently, not all trees were lopped on crop fields. For instance, 30% and 44% of farmers in Bunkure and Maigateri LGAs, respectively, predominantly allowed their Neem (*Azadirachta indica*) trees to remain on the crop fields validating the finding of (Bayala et al. 2011). About 36% (Bunkure) and 42% (Maigateri) of farmers sowed some selected trees that are used mainly as live fences around their farms with 38% of its biomass mostly used as fuel in Bunkure LGA and 31% as manure in Maigateri LGA.

Furthermore, of all the trees allowed to remain on the farm, none met the criteria for the study (i.e., dry season feed, enriching soil nitrogen, increasing protein and energy content in fodder while mitigating GHG emissions like Locust-beans (*Parkia biglobosa*), Camel's foot/*Kalgo* (*Piliostigma reticulatum*) and are distinctively the most popular leguminous tree owned by 10% in Bunkure LGA and 24% of farmers in Maigateri LGA. Personal conversation with staff members of the outstation office of Forestry Research Institute of Nigeria (FRIN), in Samaru (Zaria, Kaduna State) revealed that Locust-beans/*Dorowa* (*Parkia biglobosa*), *Kalgo* (*Piliostigma reticulatum)*, and *Gawo* (*Acacia albida*) trees were generally not planted because their seedlings or seeds were undomesticated thus, virtually unavailable. It is further compounded by the long period of growth (Bayala et al. 2014). However, efforts of FRIN were said to be underway to ensure the availability of Locust-beans (*Parkia biglobosa*) seeds or seedlings for propagation with growth period under 7 years as against the natural one of between 12 to 15 years.

Sorghum was the predominant cereal grown by over 50% of farmers in the two study locations. Most farmers in both LGAs (50% and 53%), reported that their sorghum residues comprised an adaptation plan for dry season supplementary feed/feed reserve for their ruminant livestock. About 60% of farmers in Bunkure LGA cultivated Soybean while, 53% in Maigateri LGA cultivated groundnut as their predominant legumes. Despite the numerous benefits of soybean and groundnut residue biomass for nitrogen fixation and GHG mitigation when used for soil amendments as an adaptation strategy, majority of farmers in Bunkure and Maigateri LGA sold their crop residue biomass. The implication of these practices are consequential as it may lead to more emissions of CH_4, CO_2, and N_2O, which may result in accelerated nutrient depletion from burning of bushes and crop residues, leaching, and acid rains (NH_3 of volatilization) from the application of inorganic fertilizer (IPCC 2013). Once there is accelerated nutrient depletion, there may be gross production declines.

Farmers in the study zones claimed that unplanned consumption by free ranging livestock reduced the incentives for use of crop residues soil amendments. Therefore, except by-laws and community conventions are in place to restrict free grazing, the use of biomass for soil amendment faces constraints for the poorly sourced farmers of Bunkure and Maigateri LGAs, who would rather sell their crop residue than have them consumed by free-ranging livestock.

Adaptation strategies such as change in sowing dates and use of early maturing varieties were adopted by most farmers in the regions. However, such practices in addition to mulching are merely coping/autonomous or opportunist strategies by

Table 2 Livestock production reported by households in Bunkure and Maigateri

	Bunkure		Maigateri	
	N = 50		N = 55	
Livestock characteristic	Frequency	Percentage	Frequency	Percentage
Livestock				
Predominant breed of sheep	25	50	30	55
Predominant breed of goat	32	64	40	73
Seasonal tethering/cut and carry	19	38	15	37
Intensive/peri-urban ruminant husbandry	1	2	1	2

farmers as described by Haider (Haider 2019). More is required of these practices in terms of investment in infrastructure, subsidies, research, innovation, and tax regimes to attain planned adaptation strategies.

Livestock production is a major activity among smallholder farmers in Bunkure and Maigateri, LGAs. During the focus group discussions, farmers reported that most crops were sold immediately after harvest. Revenues from crop sales were used to purchase animals, particularly ruminant livestock. The Red Bororo (100%) is the major breed of cow in Maigateri LGA, whereas Sokoto Gudali is reared by 60% of farmers in Bunkure LGA (Table 2). A cow, 5 sheep, and 6 goats are the average herd size of cattle, sheep (Uda), and goats (Red Sokoto or Maradi Breed) owned per household in Bunkure. Households in Maigateri owned an average of 2 cows, 12 sheep, and 14 goats. Livestock is fattened for up to 7 months and then sold. Part of the revenues generated from livestock sales is used to acquire farm inputs for another planting season.

Farmers in Bunkure and Maigateri LGAs believe a successful adaptation strategy is "compound dairy" (a traditional system of milking in which animals are confined to the compound rather than allowed to graze). The cows supply the family with fluid milk for consumption and any excess above household needs is sold or processed into other dairy products. In this system, dairy animals were either tethered and fed on traditional cut and carry using cereal/legume crop residue or allowed to graze freely around the compounds in the day and confined in the night in the wet season. It was the most widely practiced adaptation mechanism among all households as confirmed by 98% of households.

A key issue with the compound dairy is whether it is possible to achieve higher productivity per animal from increased grain cultivation, without increasing GHG emissions. The use of concentrates or cultivation of pastures for supplementary feeding of indigenous Zebu and Ndama breeds of cattle is uneconomical and of doubtful feasibility (Ayantunde et al. 2007). Because, there is no strong evidence of economies of scale, due to under-utilized family labor and the ability of ruminants to exploit low-value roughage, including that gathered or grazed from public lands (Mcdermott et al. 2010). Moreover, higher-productivity breeds cannot generally replace indigenous Zebu and Ndama breeds of cattle, as these breeds are not suitable for North-western Nigeria (Ayantunde et al. 2011). Although studies confirm

Table 3 Optimal farm-level tree-crop-livestock production decisions for the three scenarios in Bunkure and Maigateri LGA

Outcome or activity	Units	Bunkure				Maigateri			
		Baseline	Difference with – 10% GHG reduction	Difference with – 25% GHG reduction	Maximum GHG reduction	Baseline	Difference with – 10% GHG reduction	Difference with – 25% GHG reduction	Maximum GHG reduction
Tree									
Locust beans/ Carmel's foot	Ha	**2**	0	−0	−1	**2**	0	0	0
Crop									
Sorghum	Ha	**1**	0	0	0	**1**	0	0	0
Soybean/ groundnut	Ha	**0**	0	0	0	**3**	0	−0	−2
Animals									
Cow	Head	**0**	0	0	0	**1**	−0	−1	−1
Sheep	Head	**10**	−5	−5	−5	**5**	−1	−1	−1
Goats	Head	**5**	0	0	0	**4**	0	0	0
GHG emissions per farm	Kg CO_2eq /yr	**2943**	2649	2207	2175	**3057**	2751	2293	2124
Revenues									
Locust beans/ Carmel's foot	000 ₦/ yr	**2316**	0	−4746	−6754	**8606**	0	0	0
Sorghum	000 ₦/ yr	**8334**	0	0	0	**7430**	0	0	0
Soybean/ groundnut	000 ₦/ yr	**3004**	0	0	0	**1105**	0	−1606	−6051

Cow	000 ₦/yr	**7226**	7723	0	0	**1983**	−2937	−1145	−1145
Sheep	000 ₦/yr	**2191**	−1071	−1061	−1061	**7385**	−1955	−1955	−1955
Goats	000 ₦/yr	**1230**	0	0	0	**6165**	0	0	0
Cost of purchased inputs									
Locust beans/ Carmel's foot	000 ₦/yr	**1722**	0	−3529	−5022	**6002**	0	0	0
Sorghum	000 ₦/yr	**3493**	0	0	0	**1976**	0	0	0
Soybean/ groundnut	000 ₦/yr	**218**	0	0	0	**2499**	0	−3633	−1369

improved feeding and manual management system as mitigation strategies that reduce total emissions of CO_2, CH_4, and N_2O in the livestock sector, the extent to which this is possible requires further study to determine appropriate strategies for location-specific reductions.

Farm-Level Trade-Offs Between Income and GHG Emissions

GHG reductions of up to 26% were possible with assumed minimum levels of household consumption in Bunkure LGA (Table 3), but reduction would require reductions in farm-level revenue of more than 20%. Reductions were achieved essentially from reductions in the sales of sheep meat, locust bean pod, branch, and trunk in addition to a reduction in the sales of sheep manure (Appendix Table A3). Maximum possible decreases in GHG emissions require reduction of livestock to minimum assumed household consumption levels and a reduction in profitable locust bean production through reduction in land area by 29%. Similarly, quantities of purchased input used for locust bean production, reduced significantly for urea, seeds, and agricultural pesticides. Below 26%, the model could not attain feasible adaptation strategies because it could not meet household requirements as assumed.

In Maigateri LGA, GHG reductions of up to 30.5% could be attained with minimum levels of household consumption. However, such reductions in GHG emissions would require reductions in farm-level revenue of about 44%. Reductions in GHG emissions were accomplished primarily decreases in groundnut, cow, and sheep production and from up 100% reduction in the sales in the livestock component as well as 55.7% reduction in sales of groundnut grain in the crop component. Maximum GHG emissions reduction implied decreases in groundnut production, in addition to the reduction of livestock to minimum assumed household consumption levels.

By implication, the assessment of trade-offs that indicate potential pathways for increased agricultural productivity with fewer negative environmental effects, are usually location- and context-specific, thus difficult to generalize, categorize, and describe in contrasting socio-ecological contexts (e.g., Thornton et al. 2018). Yet, it could be futile to identify win–win production without a framework to generalize them into policy recommendations and developmental actions. The major contribution of this chapter is the establishment of a generic conceptual framework that can be used across farms of different agro-ecological zone. Particularly for comparative assessment and prioritization of policy options, that could be scaled up from local/community or farm level to regional or country levels.

However, better understanding of trade-offs is needed to attain win–win scenario (Steenwerth et al. 2014; Klapwijk et al. 2014; Kanter et al. 2016). More accurate indices and numbers could be achieved with the use of system dynamic model analyses (Sterman 1989; Kopainsky and Nicholson 2015), and could be considered as suggestion for further studies.

Conclusions

This chapter has documented the importance of agriculture and land use in the generation of GHG emissions in Nigeria, and the need for both adaptation and mitigation strategies in smallholder agriculture. The farm-level impacts provide information needed to formulate appropriate mitigation and adaptation strategies for going forward in climate-smart agriculture. A key finding from the farm-level analysis is that revenue generated decreases with GHG emissions mitigation and adaptation. Maximum possible GHG emissions reductions (while still meeting minimum household consumption needs) would require marked changes in production pattern and would lower household incomes. There appear to be no win–win adaptation strategies with current production practices and technologies that both increase revenue generation and reduce GHG. Productivity-enhancing technologies that would simultaneously reduce GHG emissions and increase revenue are needed for smallholder farms to play meaningful climate change mitigation and adaptation strategies in Nigeria without the burden of reduced incomes and greater risk. We also did not assess effects over longer periods, where the role of livestock as a key asset may imply additional negative impacts. This suggests the need for programmatic and policy actions to be taken by both national agricultural research systems and the Consultative Group for International Agricultural Research (CGIAR) to promote climate-smart agriculture in Nigeria.

Acknowledgments This study was conducted with the financial support of the Standard Assessment of Mitigation Potentials in Smallholder Farmers Livelihood System (SAMPLES), a CGIAR Research Program on Climate Change, Agriculture, and Food Security (CCAFS).

Appendix 1: Detailed Description of the Optimization Model Used to Assess Trade-Offs Between Household Welfare (Income) and Farm-Level GHG Emissions

The Farm-Level Optimization Model

The farm model (Eq. 1) assessed the potential trade-offs between GHG reductions and farm revenue with the aim to identify the impacts of possible mitigation strategies and facilitate future analysis of potential adaptation opportunities through production practices that result in both reduced emissions and increased farm incomes.

The objective function Z maximizes net household full revenue generated, given by:

$$Z = -\sum_{m=1}^{12} W_m \cdot X_m^L - \sum\nolimits_{i=1,j=1}^{i=4,j=6} P_i \cdot X_{ij}^I + \sum_{j,p=1}^{20} C_{jp} \cdot X_{jp}^P \tag{1}$$

where

W_m= Wage paid per hour of hired labor in month m

X^L_m = Hours of labor hired by the household in month m

X^I_{ij} = Purchased input i for tree or crop production activity j

P_i = Price of purchased input i per unit

C_{jp} = Unit value per kg for subproduct p of tree, crop, or livestock activity j

X^P_{jp} = Production of subproduct p of tree, crop, or livestock activity j

The objective function maximizes the total value of products derived from activities $j = 1$ to 6 locust bean and camel's foot in Bunkure and Maigateri, respectively. Represented as tree species. Branches and trunk of the Locust bean and Camel's foot trees can be used to meet household firewood requirements. Sorghum, legumes species are predominantly soybean in Bunkure and groundnut in Maigateri. While livestock includes cows, goats, and sheep production. "Full income" is specified as sum of product sold by the household. In addition to the opportunity cost of product consumed by the household less the costs of hired labor and the value of purchased inputs $i = 1$ to 4 that include fertilizer (NPK mix and urea), agricultural chemicals, manure, and seed for grains and legumes. Each of the products has subproducts that are either sold to generate revenue or used as internally generated inputs (Table A3) or used as internally generated inputs on the farm (Table A1). A total of 20 *(j, p)* combinations of activities and subproducts are represented.

Land is one of the basic farm household resource constraint and the predominant land type (upland) used for rainfed production is modeled with the following equation:

$$L \geq \sum_{j=1}^{6} L \cdot X^P_j \tag{2}$$

which shows that the total land available to the household, L, must be greater than or equal to total land used for annualized production activities j, X^P_j. Thus, for model simplification, the agricultural activities were assumed to be yearly. Land rental is excluded to avoid unbounded solutions.

A labor constraint signifies that the sum of household labor H_m and or hired labor, X^L_m available must exceed or meet total labor required for agricultural production activities, X^P_j

$$H_m + X^L_m \geq \sum_{j=1}^{7} a^L_{jm} \cdot X^P_j \tag{3}$$

The purchase of inputs used for production activities is given by the product of the activity X^P_j and the fixed coefficient a^I_{ij}

$$a^I_{ij} \cdot X^P_j \leq X^I_{ij} \tag{4}$$

Inputs produced on-farm and used for other production include tree fodder, crop residue biomass, manure, and milk. Tree fodder and crop residues are used as animal feed or soil amendment. Tree components were partitioned using formula adapted from Powell et al. (1995) and Bayala et al. (2014). Residue obtained as by-product from sorghum production were derived, using formula for harvest index adapted from Powell et al. (1995). Similarly, manure is used as fertilizer, while milk is used to feed young livestock. Formula for calculating manure production was in line with Powell et al. (1995) and Ayantunde et al. (2000). The relationships are represented by this equation:

$$\sum_{j'} a^{R}_{j',j} \cdot X^{P}_{j'} \geq \sum_{j} a^{R}_{j,j'} \cdot X^{P}_{j} \tag{5}$$

The equation above requires that the amount available from the sum of production of *j'* sources must be greater than or equal to the sum of farm-generated resource used. The model includes constraints for animal nutrient requirements: dry matter (DM, kg/day), crude protein (CP, g/day) and metabolizable energy (ME, in mega joules, MJ) FAO, http://www.fao.org/ag/againfo/themes/documents/PUB6/ P6 20. htm: that states the recommended daily ME for maintenance of sheep and goat is 5.1 and for cow producing less than 5 lit per day is 46 ME MJ/day. Recommended daily CP and DM requirement is 3% animal body weight, with average weight of 350 kg cow, 20 kg sheep, and 20 kg goat translating to 400, 32, and 32 g CP/day, respectively

The constraint in equation 6 requires the diet fed to livestock must be less than or equal to the nutritional requirements for metabolizable energy (ME), crude protein (CP), and dry matter (DM) is represent thus:

$$\sum_{j',p} a^{NS}_{j',p,n} \cdot X^{USE}_{j',p,j} \forall j \in \text{animal}, (j',p) \in \text{feeds} \geq a^{NR}_{j,n} \cdot X^{P}_{j} \tag{6}$$

which indicates that the sum of the *(j,p)* the product of feeds allocated to animals of type *j*, $X^{USE}_{j',p,j}$ and the nutrient content of those feeds, $a^{NS}_{j',p,n}$, must exceed the required $n = 3$ nutrients for animals of type *j*, $a^{NR}_{j,n} \cdot X^{P}_{j}$

Products *(j, p)* allocated to household use $X^{HH}_{j,p}$, also must meet minimum household consumption requirement $\text{HHR}_{j,p}$, as follows:

$$\text{HHR}_{j,p} \leq X^{HH}_{j,p} \tag{7}$$

A balance constraint ensures that uses of products *(j, p)* for household consumption, sales, or use in production is less than or equal to the total sources supplied:

$$X^{HH}_{j,p} + X^{SALES}_{j,p} + \sum_{j'} a^{R}_{j',j} \cdot X^{P}_{j} \leq a^{YLD}_{j,p} \cdot X^{P}_{j,p} \tag{8}$$

Table A1 Assumed structure of inputs and outputs in the tree-crop livestock production optimization model

Activity (j)	Inputs	Output	Saleable output	Consumed output
Locust bean and Camel's foot trees	Seeds, urea fertilizer, pesticides, labor, manure	Fodder, pod, pod valve, bran, branches, and trunk	Fodder, pod, pod-valve, bran	Pod, branches, and trunk
Sorghum	Seeds, urea, and NPK fertilizer, pesticides, labor, manure	Grain, fodder, bran	Fodder, bran	Grain
Soybean and groundnut	Seeds, urea, and NPK fertilizer, pesticides, labor, manure	Grain, fodder, bran, hull	Grain, fodder, bran, hull	Grain
Cow	Sorghum Stover/straw, Soyhull/groundnut residue biomass	Meat, milk, manure	Meat, milk, manure	Milk
Goat	Sorghum Stover/straw, Soyhull/groundnut residue biomass	Meat, manure	Meat, manure	Meat
Sheep	Sorghum Stover/straw, Soyhull/ groundnut residue biomass	Meat, manure	Meat, manure	Meat

An additional constraint requires that the sum of GHG emissions $a_j^{GHG}X_j^P$ generated in the predominant farm practices must equal total *GHG* emissions per farm:

$$\sum_{j=1}^{6} a_j^{GHG}X_j^P = GHG \tag{9}$$

which is used in the specification of alternative scenarios about required GHG emissions reductions per farm. For simplicity, a_j^{GHG} does not include the effects of emissions from application of lime, pre-farm operations during storage, and transportation as well as all mechanized farm operations, as these are minimal in this farming system.

The Tier 1 default methods and emission-factors (EFs) from the 2006 version of the guidelines that considered management practices with soils containing N inputs (Table A2.1) was employed. The default Carbon (C) produced is less than the estimated C harvested, and dead wood and litter stocks present in the predominant production system are at equilibrium and specified as zero. The Tier 1 default EF of 0.20 representing 20% for CO $(NH_2)_2$ was assumed for calculating annual CO_2 emissions from urea application in kg C/yr. Amounts of CO_2 equivalents were calculated by multiplying CO_2–C emissions values obtained by 44/12. Non-CO_2 emissions due to fire were estimated from the product of land area, (hectare), mass of fuel available for combustion, (tons ha $^{-1}$), combustion factor values, and emission

Table A2.1 Assumed values of GHG emissions related to land use, Bunkure and Maigateri LGAs

	Bunkure			Maigateri		
Type of emissions	Locust bean	Sorghum	Soybean	Camel's foot	Sorghum	Groundnut
	(kg CO_2eq/ha/year)	(kg CO_2eq/ha/year)	(kg CO_2eq/ha/year)	(kg CO_2eq/ha/year)	(kg CO_2eq/ha/year)	(kg CO_2eq/ha/year)
Non-CO_2 from burning	127.5	127.5	127.5	127.5	127.5	127.5
Direct N_2O soils	10.1	142.3	12.3	6.6	146.4	11.4
Indirect N_2O soils	2.3	21.6	0.0	1.1	25.6	0.0
Urea application	36.7	73.3	0.0	18.3	32.1	18.3

Table A2.2 Assumed values of GHG emissions related to animal numbers, Bunkure and Maigateri LGAs

	Bunkure			Maigateri		
Type of emissions	Cows (kg CO_2eq/TLU/year)	Sheep (kg CO_2eq/TLU/year)	Goats (kg CO_2eq/TLU/year)	Cows (kg CO_2eq/TLU/year)	Sheep (kg CO_2eq/TLU/year)	Goats (kg CO_2eq/TLU/year)
CH_4 emissions	987.0	172.2	109.6	987.0	172.2	109.6
N_2O emissions	186.7	20.8	24.4	186.7	20.8	24.4

Table A3 Optimal farm-level tree-crop-livestock production decisions for the three scenarios in Bunkure and Maigateri LGA

Outcome or activity	Units	Bunkure				Maigateri			
		Baseline	Difference with – 10% GHG reduction	Difference with – 25% GHG reduction	Maximum GHG reduction	Baseline	Difference with – 10% GHG reduction	Difference with – 25% GHG reduction	Maximum GHG reduction
Sales of products	Kg/yr								
Tree									
Locust beans/ Camel's foot pod		**3383**	0	−697	−993	**1180**	0	0	0
Locust beans/ Camel's foot fodder		**0**	0	284	0	**0**	372	372	0
Locust beans/ Camel's foot branch		**3045**	0	−1073	−1527	**0**	0	0	0
Locust beans/ Camel's foot trunk		**1361**	0	−2790	−3970	**4745**	0	0	0
Crop									
Sorghum bran		**0**	0	66	0	**0**	31	31	0
Soybean/ groundnut grain		**0**	0	0	0	**2601**	0	−385	0
Soybean bran/ groundnut meal		**0**	0	3	0	**0**	13	11	0

(continued)

Table A3 (continued)

Outcome or activity	Units	Bunkure				Maigateri			
		Baseline	Difference with – 10% GHG reduction	Difference with – 25% GHG reduction	Maximum GHG reduction	Baseline	Difference with – 10% GHG reduction	Difference with – 25% GHG reduction	Maximum GHG reduction
Soyhull/ groundnut haulms		**1**	0	0	0	**0**	104	804	0
Animals									
Cow meat		**43**	46	0	0	**160**	–24	–93	–93
Cow milk		**0**	146	0	0	**187**	–48	–187	–187
Cow manure		**257**	275	0	0	**811**	–120	–468	–468
Sheep meat		**142**	–131	–142	–142	**36**	–36	–36	–36
Sheep manure		**983**	–481	–521	–521	**503**	–133	–133	–133
Goat manure		**555**	0	0	0	**432**	0	0	0

factor (g kg $^{-1}$ dry matter burnt) for various type of burning. The carbon dioxide equivalents (CO_2eq) factor was set at zero based on the assumption that the conversion of CO and NO_x has weak global warming potential.

N_2O emissions include volatilization and leaching from manure management system (MMS) and managed soil (MS). Other N_2O losses considered are annual amount of N in crop residues (above and below ground), including N-fixing crops, and from forage/pasture renewal, returned to soils annually, (kg N yr.$^{-1}$). DAP composition for Nitrogen (N) content in 50 kg bag of NPK fertilizer with urea excluded is 0.20 kg N. All emissions were converted to N_2O by multiplying 44/28 with values obtained

In ruminant livestock production, MMS and enteric fermentation result in CH_4 emissions in (Table A2.2). The EFs for sheep, goat, and mature cow grazing on large areas in developing countries were used. One TLU represents 250 kg live weight; equivalent of 1 camel, 1.43 cattle, 10 sheep/goats according to Ayantunde et al. (2011). All results were converted to carbon-dioxide equivalents (CO_2eq). In accordance with IPCC, a unit of CO_2, CH_4, and N_2O represents 1, 21, and 310 units of CO_2eq, respectively

To implement scenarios requiring farms to reduce GHG emissions from current levels, an additional constraint was developed to ensure GHG emissions generated from the predominant production practice is less than the assumed amount:

$$\sum_{j=1}^{6} a_j^{GHG} X_j^P \leq GHGLIM \tag{10}$$

References

Ajetomobi J, Olusanya A, Gbadegesin A (2015) AGRODEP working paper 0016 the potential impact of Climate Change on Nigerian agriculture. International Food Policy Research Institute, Washington, DC. https://doi.org/http://ebrary.ifpri.org/cdm/ref/collection/p15738coll2/130099

Ayantunde AA, Fernandez-Rivera S, Hiernaux PHY, van Keulen H, Udo HMJ, Chanono M (2000) Effect of nocturnal grazing and supplementation on diet selection, eating time, forage intake and weight changes of cattle. Anim Sci 71:333–340

Ayantunde AA, Delfosse P, Fernandez-Rivera S, Gerard B, Dan-Gomma A (2007) Supplementation with groundnut haulms for sheep fattening in the West African Sahel. Trop Anim Health Prod. https://doi.org/10.1007/s11250-007-9009-1

Ayantunde AA, Leeuw JD, Turner MD, Said M (2011) Challenges of assessing the sustainability of (agro) -pastoral systems. Livest Sci 139(1–2):30–43. https://doi.org/10.1016/j.livsci.2011.03.019

Ayinde TB (2019) Economics of optimum tree-crop-livestock intensification and green house gas emissions in smallholder production system in North-western Nigeria. Unpublished Ph.D. thesis at the Department of Agricultural Economics, Ahmadu Bello University, Zaria

Bayala J, Kalinganire A, Tchoundjeu Z, Sinclair F, Garrity D (2011) Conservation agriculture with trees in the West African Sahel – a review (No. 14). ICRAF, World Agroforestry Centre, Nairobi

Bayala J, Ky-dembele C, Kalinganire A, Olivier A, Nantoumé H (2014) A review of pasture and fodder production and productivity for small ruminants in the Sahel. World Agro Forestry Center, Bamako. Retrieved from www.worldagroforestry.org

Change Climate Response Policy (2015) Nigeria's intended nationally determined contribution 1 summary. Federal Republic of Nigeria

Dunkelberg E, Finkbeiner M, Hirschl B (2014) ScienceDirect sugarcane ethanol production in Malawi: measures to optimize the carbon footprint and to avoid indirect emissions. Biomass Bioenergy 71:37–45. https://doi.org/10.1016/j.biombioe.2013.10.006

Dupriez H, De Leener P (1998) Tree and multistorey agriculture in Africa. Technical Centre for Agricultural and Rural Cooperation(CTA)

Federal Republic of Nigeria (2018) Federal Republic of Nigeria First Biennial Update Report (BUR1) of the Federal Republic of Nigeria under the United Nations Framework Convention on First Biennial Update Report (BUR1) of the Federal Republic of Nigeria. https://doi.org/https://unfccc.int/sites/default/files/resource/Nigeria%20BUR1_Final%20%282%29.pdf

Haider H (2019) Climate change in Nigeria: impacts and responses. UK Department for International Development (DFID), K4D and other government departments

Hansen JW, Araba D, Hellin J, Goslinga R (2017) A roadmap for evidence-based insurance development for Nigeria's farmers. Wageningen: CGIAR research program on climate change, agriculture and food security (CCAFS). Available online at: www.ccafs.cgiar.org

Inter-Governmental Panel on Climate Change (IPCC) (2007) In: 4th assessment report. Impacts adaptation and vulnerability. Cambridge

Inter-Governmental Panel on Climate Change (IPCC) (2013) Summary for policymakers. In: Stocker TF, Qin D, Plattner GK, Tignor MMB, Allen SK, Boschung J, Nauels A, Xia Y, Bex V, Midgley PM (eds) Climate Change 2013: the physical science basis. Cambridge University Press, Cambridge/New York

Kanter DR, Musumba M, Wood SLR, Palm C, Antle J, Balvanera P et al (2016) Evaluating agricultural trade-offs in the age of sustainable development. Agric Syst. (in Press)

Klapwijk C, van Wijk M, Rosenstock T, van Asten P, Thornton P, Giller K (2014) Analysis of trade-offs in agricultural systems: current status and way forward. Curr Opin Environ Sustain 6:110–115

Kopainsky B, Nicholson FC (2015) System dynamics and sustainable intensification of food systems: complementarities and challenges. In: The 33rd international conference of the system dynamics society, 19–23 July 2015. Cambridge, MA, pp 15–22

Mcdermott JJ, Staal SJ, Freeman HA, Herrero M, Van De Steeg JA (2010) Sustaining intensification of smallholder livestock systems in the tropics. Livest Sci 130:95–109. https://doi.org/10.1016/j.livsci.2010.02.014

National Agricultural Extension and Research Liaison Services (NAERLS), & Planning Research and Statistic Department (PRSD). (2012). In: Agricultural Performance Survey Report of 2012 Wet Season in Nigeria, NAERLS, Ahmadu Bello University. (In: Agricu, p. 222.). Zaria, Nigeria: NAERLS Press

Nicholson CF, Tedeschi LO, Lellis Vieira ACG (2011) The application of system dynamics modelling to enhance profitability and sustainability in Latin American livestock systems. In: Paper presented at the II Simposio Internacional Genómica y Modelación en los Nuevos Escenarios de la Ganadería Bovina Tropical. Palmira

Powell JM, Fernandez-Rivera S, William TO, Renard C (1995) Livestock and sustainable nutrient cycling in mixed farming Systems of Saharan Africa. In: Powell C, Fernandez-Rivera JM, William S, Renard TO (eds) Proceedings of an international conference, Addis Ababa, p 568

Steenwerth KL, Hodson AK, Bloom AJ, Carter MR, Cattaneo A, Chartres CJ, …, Jackson LE (2014) Climate-smart agriculture global research agenda: scientific basis for action. Agric Food Sec 3(1):11

Sterman JD (1989) Modeling managerial behavior: misperceptions of feedback in a dynamic decision-making experiment. Manag Sci 35(3):321–339

Thornton PK, Whitbread A, Baedeker T, Cairns J, Claessens L, Baethgen W et al (2018) A framework for priority-setting in climate smart agriculture research. Agric Syst 167(October):161–175. https://doi.org/10.1016/j.agsy.2018.09.009

United Nations Department of Economic and Social Affairs: Population Division (2020) Nigeria population clock (live). https://doi.org/https://countrymeters.info/en/Nigeria

United States Aid for International Development (USAID) (2019) Greenhouse gas emissions in Nigeria

UN-REDD Programme (2015) Draft report National Programme Annual Report Federal Republic of Nigeria. UN-REDD programme

World Bank (2020) Nigeria – agricultural land (hectares). https://doi.org/https://tradingeconomics.com/nigeria/agricultural-land-hectares-wb-data.html

2

Socioeconomically Informed Use of Geostatistics to Track Adaptation of Resource-Poor Communities to Climate Change

Martin Munashe Chari, Hamisai Hamandawana and Leocadia Zhou

Contents

M. M. Chari (✉)
Department of Geography and Environmental Science, University of Fort Hare, Alice, South Africa

Risk and Vulnerability Science Centre (RVSC), University of Fort Hare, Alice, South Africa
e-mail: martinmchari@gmail.com

H. Hamandawana
Department of Geographical Information Systems (GIS) and Remote Sensing, University of Fort Hare, Alice, South Africa
e-mail: hamandawanah@yahoo.com

L. Zhou
Risk and Vulnerability Science Centre (RVSC), University of Fort Hare, Alice, South Africa
e-mail: Lzhou@ufh.ac.za; zhouleocadia@gmail.com

Abstract

As the Green Climate Fund continues to make concerted efforts to leverage funding for resource-constrained communities in the global south under the aegis of increasing climate change impacts in sub-Saharan Africa, there is urgent and compelling need for tools that assist organizations to track the effectiveness of adaptation interventions in reducing vulnerability. This chapter offers a cost-effective methodology to track adaptation by using a case-study-based identification of communities with diminishing coping capacities in Raymond Mhlaba Local Municipality in the Eastern Cape Province of South Africa. Multistep geostatistical techniques were utilized in the ArcGIS 10.5 software environment to rank and spatialize changes in adaptation by using demographic census data for the years 2001 and 2011. Results of the analysis revealed that 12 communities had declining or static adaptive capacities between 2001 and 2011, while 10 communities had long-term decrease in adaptive capacities from 2001 to 2011 from a sampling universe of 134 communities. These findings are important because they demonstrate that the methodology can be effectively used to provide actionable information on the prevalence of low adaptation capacities at appropriate temporal and spatial scales, in order to guide the allocation of limited resources to the most deserving communities.

Keywords

Climate change · Adaptation tracking · Geostatistics · South Africa

Introduction

Africa is one of the most vulnerable continents to climate change (Sarkodie and Strezov 2019; IPCC 2014; World Bank 2014) due to high dependence of livelihoods on climate-sensitive sectors such as agriculture and forestry, poor infrastructure, and limited adaptive capacities to cope with adverse impacts (Ford et al. 2015). In an effort to overcome these constraints, the Paris Agreement and the Katowice Climate Package have been urging all parties to take more climate action by encouraging the adoption and documentation of appropriate adaptation strategies. Adaptation has been getting increasing recognition as a fundamental component of global-level climate change strategies and investigations (Leal Filho and de Freitas 2018; Brooks 2011; Biesbroek et al. 2010), with sub-Saharan Africa emerging in the limelight (Fox et al. 2018; Jiri and Mafongoya 2017; Leal Filho et al. 2015; Leal Filho et al. 2017; Ziervogel et al. 2014; Dinar et al. 2012) because of growing apprehension of climate risks. This apprehension explains why the Green Climate Fund (GCF) has been increasing levels of adaptation funding to many countries in Africa in order to fulfill the Paris Agreement (Pauw et al. 2018; Tompkins et al. 2018; Georgeson et al. 2016) by enhancing their capacities to mitigate the adverse effects of climate change (Mukarakate 2016). This commitment is demonstrated by the recent extension of support to as many as four climate change adaptation projects in South Africa alone (https://www.greenclimate.fund/countries/south-africa). The GCF is part-implementation of the Paris Agreement in the context that "Developed country Parties shall provide financial resources to assist developing country Parties with respect to both mitigation and adaptation in continuation of their existing obligations under the Convention." Because adaptation funding is increasing, tracking progress in its effectiveness is essential in order to identify constraints that need to be addressed and successes that require further strengthening and improvement.

Adaptation tracking is a part of intervention monitoring and assessment that helps to capture the effectiveness with which coping strategies are translated into tangible courses of action that reduce vulnerability to climate change (Berrang-Ford et al. 2019; Lesnikowski et al. 2017). It is also useful in the governing of adaptation actions by providing a baseline for continuous monitoring and evaluation of progress over time (FAO 2017). Although necessity to monitor the advancement of climate change adaptation is gaining increasing recognition, effective tracking continues to be undermined by lack of objective indicators for scoping how adaptation takes place (Ford et al. 2013). This challenge is aggravated by difficulties in quantifying the expression of adaptation tracking due to lack of reliable tools that can be used to identify trends and gaps in adaptation responses. With climate projections pointing to relatively strong rates of warming over Africa (Engelbrecht et al. 2015) and global temperatures rising by an estimated 1.5 °C above preindustrial levels, it is apparent that there is immediate need to meaningfully embrace climate friendly adaptation strategies (IPCC 2018; New 2018). As atmospheric greenhouse gas concentrations increase in South Africa, an analysis of climate trends, using observations between 1980 and 2016 and IPCC-validated model-simulations up to 2050, projected an increase in global temperatures by 0.02 °C/year up to 2050, and a possible increase to 0.03 °C/year in future (Jury 2019). These scenarios have severe implications for South Africa's climatically vulnerable communities, and there is convincing evidence to support the view that the need to formulate innovative techniques that can be used to monitor the translation of adaptation plans into implementable interventions is long overdue.

In South Africa, Limpopo and Kwa-Zulu Natal provinces are extremely vulnerable to climate change-related problems due to their high dependence on climate-sensitive sources of livelihood (Rankoana 2019; Hlahla et al. 2018; Ncube et al. 2016; Gbetibouo et al. 2010). Out of the country's nine provinces, the Eastern Cape is recognized as the most vulnerable (Zhou et al. 2016) because of its susceptibility to consecutive droughts (Ngqakamba 2019; Ndamase 2019; ADM 2010, 2012, 2017; IFRC 2004) with one of the most severe outbreaks being experienced in Amathole District Municipality's (ADM) Raymond Mhlaba Local Municipality (RMLM) during the 2018/19 season (Ndamase 2019). The severity of this drought is demonstrated by four of the six local municipalities under ADM's jurisdiction namely: RMLM, Mbhashe, Mnquma, Ngqushwa that received two water trucks each, while the remaining two; namely Great Kei and Amahlathi, received one truck each (Dwesini 2018). Despite the increasing incidence of droughts in RMLM, the implementation of effective climate change adaptation strategies is still ineffective and continues to be undermined by conspicuous absence of reliable adaptation tracking techniques (ADM 2017). These bottlenecks and challenges justify the need for a methodology that can be used to effectively track adaptation. Although demographic indicators provided in numerical format are useful measures of adaptation (Gamble et al. 2013; van Aalst et al. 2008; Wall and Marzall 2006), they are often inadequately exploited because they require a lot of computational manipulations before they can be translated into spatially intelligible information which can be used to direct attention to areas in need of support (Qiu et al. 2019; de Sherbinin 2016; Schensul et al. 2013).

This limitation is aggravated by the fact that most national-level reports on adaptation tend to present critical information in the form of spatially disjointed metrics that can be used more effectively by presenting them in coherent formats that are capable of directing practitioners to specific target areas when action is required. A structured consolidation of these fragmented regional scale observations into discrete geographical localities is therefore essential and helpful because it provides for bottom-up scientific investigations in which the local informs the regional by facilitating exhaustive interrogation of cause-and-effect relationships and identification of areas where action is needed (Hamandawana et al. 2008). The strength of this approach is demonstrated by the Paris Agreement's adoption of an inclusive agenda on what countries can do with climate change (Conway et al. 2019; Kuwornu 2019) by using grassroots strategies that promote effective implementation of adaptation strategies (Keskitalo and Preston 2019). Mindful appreciation of this inclusiveness is demonstrated by Chari et al. (2018); Bouroncle et al. (2017); Weis et al. (2016); and Juhola and Kruse (2015) who provide useful examples of approaches that can be used to promote adaptation by aiding objective identification of communities that are vulnerable to the adverse effects of climate change.

Although RMLM has not been very successful in attracting support for adaptation initiatives, evidence suggests that local communities in this area are increasingly being exposed to climate-driven short-term variabilities that are considered to be more important stimuli to adaptive responses than long-term changes in climate (Berrang-Ford et al. 2011). Unfortunately, it has not been possible for resource-poor communities in this area and others elsewhere to access meaningful assistance due to lack of local-level information that can be used to delimit localities in need of external support (Taylor 2016). This chapter attempts to demonstrate how this gap can be bridged by providing an adaptable and spatially explicit case-study-based methodology to track the extent to which local communities in RMLM have been able to meaningfully embrace climate-friendly adaptation.

Materials and Methods

Study Area

Raymond Mhlaba (Fig. 1) is a sparsely populated (~24 people/km^2) countryside local municipality comprising 23 wards that cover approximately 6,358km^2 (ADM 2017).

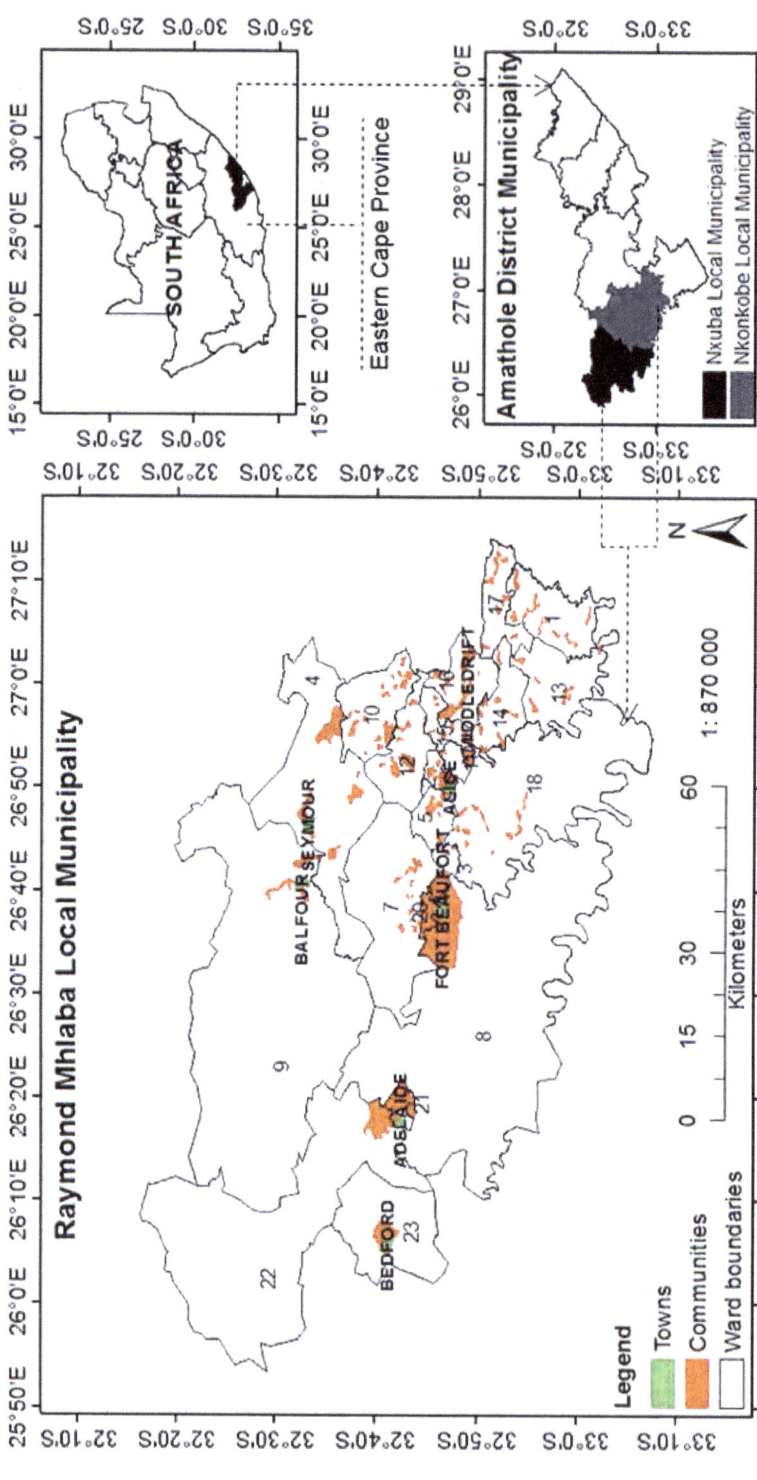

Fig. 1 Location of Raymond Mhlaba Local Municipality. (Source: Chari 2020)

It became the largest of ADM's six local municipalities when it was established by merging Nkonkobe and Nxuba Local Municipalities after the August 2016 local elections (Local Government Handbook 2016). The majority of villages in this area are situated in the immediate peripheries of Fort Beaufort (Fig. 1) with this positioning being largely influenced by accessibility to the latter's urban amenities and availability of arable land for subsistence farming. The entire municipality has a semiarid climate that is characterized by a) unreliable seasonal autumn rainfall which does not exceed 600 mm/annum with the lowest (~7 mm) and highest (~66 mm) amounts occurring in July and March, respectively, and b) average mid-day temperatures that range from 19.3 °C in June to 28.3 °C in January.

The abilities of this area's local communities to adapt to deteriorating climatic conditions are severely undermined by unreliable rainfall which makes rain-fed subsistence crop production extremely risky. This limitation and the municipality's documented failure to attract substantial adaptation funding and its positioning in a hotspot area with resource-poor households that are projected to experience widespread climate-change induced decrease in crop yields (Ncube et al. 2016) make it a suitable target for a case study-based exploration of novel techniques that can be used to boost adaptation tracking.

Data

The datasets that were used in this study were compiled by sourcing information from different sources over different time periods between January 2017 and May 2019. These datasets include (a) a shapefile that captures the distributions of local communities in RMLM as compiled by Statistics South Africa (StatsSA) in 2011, (b) 2001 and 2011 census data in a raw Microsoft Excel table format that was provided by the same source, and (c) insights of expert informants that were used to verify the reliability and appropriateness of the municipal-level data that was used in this study. Availability of census data for the years 2001 and 2011 was judged to be timely because these records provided demographic indicators over a 10-year period which was reasoned to be long enough for individual households to translate climate-change-induced coping strategies into quantifiable indicators of adaptation. These datasets were also considered to be reliable because in South Africa, StatsSA is the custodian of official multilevel planning statistics (www.statssa.gov.za).

Methods

A hybrid approach comprising a multistep GIS-based mapping and analysis of aggregated indicators was used to track the adaptation of resource-poor communities to climate change with adaptive capacity indicators being purposefully selected to accommodate representative inclusion of different communities and the singular geographical and socioeconomic distinctions of their localities. The Nkonkobe Integrated Development Plan (IDP) of 2012–2017 was used to aid the identification of communities with different livelihood activities because most of the communities in RMLM reside in the former Nkonkobe Local Municipality (Fig. 1). The selection of assessment indicators was based on the definition of adaptive capacity provided by Heltberg and Bonch-Osmolovskiy (2011) and the type and level of demographic data available for the municipality with further refinements being made by using ancillary information that was solicited from expert informants. Overall, the indicators listed and described in Table 1 below were finally selected on the basis of the logic described above and used to assess and map adaptive capacities in the municipality.

Prior to analysis, spreadsheets with information on sub-places/communities were cleaned by (1) deleting unnecessary details from all Excel datasets and (2) selecting matching names in the 2001 and 2011 census records to facilitate the execution of

Table 1 Description of demographic indicators used for assessing and tracking adaptation

Narrative indicator	Rationale	Ranking/classification of communities
Literacy levels	Specified adaptive capacity within the communities on basis of highest education level. *Reclassified from 21 classes to 6 classes.*	Scores of 0 to 5 were allocated as follows on the basis of level of education: No schooling = 0, Some Primary = 1, Completed Primary = 2, Some Secondary = 3, Completed Secondary = 4, Higher = 5.
Income levels	Communities with low incomes have low resilience because they are unlikely to have access to credit. *Reclassified from 12 classes to 5 classes.*	Scores of 0 to 4 were allocated depending on income profiles. Wards with the lowest incomes were allocated a score of 0. Groups were segregated according to annual income as follows: No income = 0, R1 – R9 600 = 1, R9 601 – R19 200 = 2, R19 201 – R38 400 = 3, ≥ R38 401 = 4.
Age profiles	Used to identify communities with low resilience to climate change by virtue of being economically inactive due to predominance of children and/ or old people in the population. *Reclassified from 18 classes to 4 classes.*	Scores of 0 to 3 were allocated and communities segregated into three age groups as follows: 0–14: Child (score of 0), 15–39: Young (score of 2), 40–59: Old (score of 3): *Senior citizen 60+ (score of 1).
		* *This categorization is based on age criteria used by South Africa Social Security Association (SASSA) to identify beneficiaries of social security assistance*
Water access by source type	Specified adaptive capacities of individual communities on the basis of most used water sources. *Reclassified from 7 classes to 4 classes.*	Other sources = 0, Surface water = 1, Ground water = 2, Regional water scheme = 3.

Source: Adapted from Chari et al. (2018)

valid geospatial analysis operations. To avoid join operation errors in ArcGIS 10.5, all communities that were founded after year 2001 and all extra spaces in the attribute tables of the community names field (key fields) were omitted from the spreadsheet of 2011. Thereafter, shapefiles were created by linking each input indicator's Excel table to the attribute table of the digitized-communities shapefiles and the Join operation used to translate these datasets into spatial layers that were combined with other layers to provide thematic map-portrayals of adaptive capacity ranks. The demographic data from StatsSA was analyzed at community level because this was the lowest level at which required data was available. The combination of layers was followed by turning off repeated fields in the output attribute tables in order to mute field redundancies when the data was exported to the geodatabase. In the final step the following Python script (Script 1) was used to automatically assign scores to communities in the income levels attribute table, and the three indicators (Literacy levels, Source of water, Age profiles) for census years 2001 and 2011 after changing their table names, field names, and row numbers, following the criteria in Table 1.

Script 1 Script That was Used to Automatically Assign Scores for Community Income Levels

```
##Name of script: Income2001.py
##Purpose: Automated allocation of scores to communities for the
2001 income levels data
#Importing system modules
import arcpy
#Input data and input fields and their lengths
table = "E:/GIS/RayMhlaba.mdb/Income_2001"
fields = ["No_income", "R1_R9600", "R9601_R19200", "R19201_R38400",
"R38401_more"]
#Adding fields to input table to store maximum and field names
maxfield = "HIGHEST"
maxname = "HIGHEST_name"
scores = "In2001_score"
arcpy.AddField_management(table, maxfield, "TEXT")
arcpy.AddField_management(table, maxname, "TEXT")
arcpy.AddField_management(table, scores, "SHORT")
#Adding created fields to the array
fields2 = fields[:] # Shallow copy
fields2.extend([maxfield, maxname, scores])
#Checking and updating of fields
with arcpy.da.UpdateCursor(table, fields2) as cursor:
  for row in cursor:
    arrayVals = [row[0], row[1], row[2], row[3], row[4]]
    highest = max(arrayVals)
    row[5] = highest
    row[6] = fields[arrayVals.index(highest)]
    if row[6] == "No_income":
        row[7] = 0
        cursor.updateRow(row)
    elif row[6] == "R1_R9600":
        row[7] = 1
```

```
        cursor.updateRow(row)
    elif row[6] == "R9601_R19200":
        row[7] = 2
        cursor.updateRow(row)
    elif row[6] == "R19201_R38400":
        row[7] = 3
        cursor.updateRow(row)
    else:
        row[7]= 4
        cursor.updateRow(row)
["No_income","R1_R9600","R9601_R19200","R19201_R38400",
"R38401_more"]
py.AddField_management(table,maxfield,"TEXT")
py.AddField_management(table,maxname,"TEXT")
AddField_management(table,scores,"SHORT")
extend([maxfield,maxname,scores])
arcpy.da.UpdateCursor(table,fields2) as cursor:
    cursor:
    arrayVals = [row[0],row[1],row[2],row[3],row[4]]
    highest = max(arrayVals)
    row [5] = highest
row[6] = fields[arrayVals.index(highest)]
    if row[6] == "No_income":
    row[7] = 0
cursor.updateRow(row)
    elif row [6] == "R1_R9600":
    row[7] = 1
cursor.updateRow(row)
    elif row [6] == "R9601_R19200":
    row[7] = 2
cursor.updateRow(row)
    elif row[6] == "R19201_R38400":
    row[7] = 3
cursor.updateRow(row)
    else:
    row[7] = 4
cursor.updateRow(row)
```

Script 1 was executed by using the Python `execfile` command to generate a map for each of the four indicators in ArcMap 10.5 for years 2001 and 2011 on the basis of the assigned scores in order to reveal spatial variations in each indicator with communities that had the lowest scores for each indicator being identified by examining all attribute tables. Adaptive capacity scores for both census years (Table 2) were automatically determined by generating a new shapefile and attribute table into which the previously calculated scores from the four indicators for each community were imported and automatically summed up by using Script 2 and Script 3. Although all indicators were weighted equally because of limited availability of indicator data for the municipality, this relaxation did not compromise the reliability of results because information obtained from expert informants and the Nkonkobe IDP for 2012–2017 (Nkonkobe Local Municipality 2012) confirmed that

Table 2 Evaluation of adaptive capacity in the attribute table for each of the census years

Village	Ward number	Access to water	Literacy levels	Income levels	Age profile	Adaptive score	Adaptive capacity
V_1	W_1	3	5	4	3	15	HIGH
V_2	W_2	S_1	S_2	S_3	S_4	.	.
.	.	.	.	.	.	.	.
.	.	.	.	.	.	.	.
V_{134}	W_{23}	3	3	2	0	8	MEDIUM

$V_1 \ldots V_{134}$ – community names; W – ward numbers; S_1, S_2, S_3, S_4 – indicator scores. Adaptive scores ranged from 1–15

Source: Adapted from Chari et al. (2018)

all indicators were equally important for adaptive capacity assessment. Adaptive capacities for the census years 2001 and 2011 were calculated in the attribute table (Table 2) by using the following formula:

$$\text{Adaptive score} = \left(\sum_{n=s1}^{s4} n\right) \tag{1}$$

Where S_1, S_2, S_3, S_4 are scores for each of the four indicators

The following Python script (Script 2) was used to automatically create a new shapefile and attribute table and to join, Age_Profile scores to the IndicatorScores table. The same script was also used to join the remaining three indicator scores to the IndicatorScores attribute table by using community name as the linking field.

Script 2 Script That was Used for Automated Creation of a New Shapefile and Attribute Table and Joining of Age_Profile Scores to the IndicatorScores Table

```
##Name of script: ScoreInput2001.py
##Purpose: Automated creation of a new shapefile and attribute table
for saving an integration of the four indicator scores and
automated joining of the remaining 3 indicator scores to the
IndicatorScores table using MP_NAME as the linking field.
#Importing system modules
import arcpy
from arcpy import env
#Setting the environment
env.workspace = "E:/GIS/"
# Specifying the input feature class, output location and feature
classes
inFeatures = "E:/GIS/RayMhlaba.mdb/Age_2001"
outLocation = "E:/GIS/RayMhlaba.mdb"
outFeatureClass = "Ind_Scores2001"
# Listing fields to be retained
myfields = ["MP_NAME", "A2001_score"]
# Creating an empty field mapping object
mapS = arcpy.FieldMappings()
# Creating an individual field map each field, and adding it to the
field mapping object
for field in myfields:
```

```
    map = arcpy.FieldMap()
    map.addInputField(inFeatures, field)
    mapS.addFieldMap(map)
    # Copying the feature class using the fields
    arcpy.FeatureClassToFeatureClass_conversion (inFeatures,
    outLocation, outFeatureClass, field_mapping=mapS)
#Joining the remaining 3 indicator scores for different fields into
one table
arcpy.JoinField_management("Ind_Scores2001", "MP_NAME",
"Income_2001", "MP_NAME", "In2001_score")
arcpy.JoinField_management("Ind_Scores2001", "MP_NAME",
"Literacy_2001", "MP_NAME", "Lit2001_score")
arcpy.JoinField_management("Ind_Scores2001", "MP_NAME",
"Water_2001", "MP_NAME", "W2001_score")
    arcpy import env
    #Setting the environment
env.workspace = "E:/GIS/"
    # Specifying the input feature class, output location and
    feature classes
    inFeatures = "E:/GIS/RayMhlaba.mdb/Age_2001"
    outLocation = "E:/GIS/RayMhlaba.mdb"
    outFeatureClass = "Ind_Scores2001"
    # Listing fields to be retained
myfields = ["MP_NAME", "A2001_score"]
    # Creating an empty field mapping object
mapS = arcpy.FieldMappings()
    # Creating an individual field map each field, and adding it to
    the field mapping object
    for field in myfields:
map = arcpy.FieldMap()
map.addInputField(inFeatures,field)
mapS.addFieldMap(map)
    # Copying the feature class using the fields
arcpy.FeatureClassToFeatureClass_conversion (inFeatures,
outLocation,outFeatureClass,field_mapping=mapS)
    #Joining the remaining 3 indicator scores for different fields
    into one table
arcpy.JoinField_management
("Ind_Scores2001","MP_NAME","Income_2001","MP_NAME",
"In2001_score")
arcpy.JoinField_management
("Ind_Scores2001","MP_NAME","Literacy_2001","MP_NAME",
"Lit2001_score")
arcpy.JoinField_management
("Ind_Scores2001","MP_NAME","Water_2001","MP_NAME","W2001_score")
```

Since the highest attainable adaptive capacity score from addition of the four highest indicator scores was 15 (Table 1), the computed adaptive capacity scores were ranked into low-medium-high adaptive capacity as follows: 1–5 = LOW, 6–10 = MEDIUM, and 11–15 = HIGH by using the Python `execfile` command to run the following Python script (Script 3) which automatically added a field for the ranked adaptive capacity scores attribute table.

Script 3 Script That Was Used for Joining of the Remaining 3 Indicator Scores to the IndicatorsScores Table Using MP_NAME as the Linking Field

```
## Name of script: AD2001.py
## Purpose: Automated summation and ranking of the four indicator
scores for year ## 2001
#Importing system modules
import arcpy, math
#Importing scores from indicator attribute tables into the Adaptive
capacity shapefile
table = "E:/GIS/RayMhlaba.mdb/Ind_Scores2001"
fields = ["A2001_score","In2001_score","Lit2001_score",
"W2001_score"]
# Adding fields to input table to store maximum and field name
total = "AD2001_Score"
rating = "ACRating_2001"
arcpy.AddField_management(table, total, "SHORT")
arcpy.AddField_management(table, rating, "TEXT")
#Adding created fields to the array
fields2 = fields[:]
fields2.extend([total, rating])
#Classifying community-level adaptive capacity scores
with arcpy.da.UpdateCursor(table, fields2) as cursor:
   for row in cursor:
      arrayVals = [row[0], row[1], row[2], row[3]]
#Calculating adaptive capacity for each community by summing the 4
indicator scores
   summation = sum(arrayVals)
   row[4] = summation
   #Allocating the adaptive capacity rating
   if row[4] <=5:
      row[5] = 'LOW'
      cursor.updateRow(row)
   elif row[4]> 5 and row[4]<=10:
      row[5] = 'MEDIUM'
      cursor.updateRow(row)
   else:
      row[5] = 'HIGH'
      cursor.updateRow(row)
      ## Purpose: Automated summation and ranking of the four
      indicator scores for year ## 2001
      #Importing system modules
import arcpy,math
   #Importing scores from indicator attribute tables into the
   Adaptive capacity shapefile
   table = "E:/GIS/RayMhlaba.mdb/Ind_Scores2001"
fields = ["A2001_score","In2001_score","Lit2001_score",
"W2001_score"]
   # Adding fields to input table to store maximum and field name
   total = "AD2001_Score"
   rating = "ACRating_2001"
arcpy.AddField_management(table,total,"SHORT")
arcpy.AddField_management(table,rating,"TEXT")
   #Adding created fields to the array
```

```
    fields2 = fields [:]
fields2.extend([total,rating])
    #Classifying community-level adaptive capacity scores
with arcpy.da.UpdateCursor(table,fields2) as cursor:
    for row in cursor :
    arrayVals = [row[0],row[1],row[2],row[3]]
    #Calculating adaptive capacity for each community by summing
    the 4 indicator scores
    summation = sum (arrayVals)
    row [4] = summation
    #Allocating the adaptive capacity rating
    if row [4] <=5:
    row [5] = 'LOW'
cursor.updateRow(row)
    elif row [4]>5androw[4]<=10:
    row [5] = 'MEDIUM'
cursor.updateRow(row)
    else: row[5] = 'HIGH'
cursor.updateRow(row)
```

An adaptive capacity map was generated in ArcMap 10.5 from the rankings that were obtained for each of the two census years and the following Python script (Script 4) subsequently used to automatically create a new shapefile and to save adaptive capacities from the indicator scores tables for years 2001 and 2011 in one attribute table (Table 3).

Script 4 Script That Was Used for Joining of the Remaining 3 Indicator Scores to the IndicatorsScores Table Using MP_NAME as the Linking Field

```
## Name of script: AD_Change.py
## Purpose: Automated creation of a new shapefile and saving
adaptive capacities for years 2001 and 2011 into one attribute
table
#Importing system modules
import arcpy
from arcpy import env
```

Table 3 Evaluation of changes in adaptive capacity scores between census years 2001 and 2011

Village	Ward number		Adaptive score			Difference	Rating
			2011	2001			
V_1	W_1		5	8		−3	DECREASE
V_2	W_2		A_1	A_2		.	–
.	.		.	.		.	.
.	.		.	.		.	.
V_{134}	W_{23}		12	8		4	INCREASE

$V_1 \ldots V_{134}$ – community names; W – ward numbers; A_1, A_2 – adaptive capacity scores for year 2011 and year 2001, respectively
Source: Chari 2020

```
#Setting the environment
env.workspace = "E:/GIS/"
# Specifying of input feature class, output location and feature
classes
inFeatures = "E:/GIS/RayMhlaba.mdb/Ind_Scores2011"
outLocation = "E:/GIS/RayMhlaba.mdb"
outFeatureClass = "AD_Diff"
# Listing of fields to be retained
myfields = ["MP_NAME", "AD2011_Score"]
# Creating an empty field mapping object
mapS = arcpy.FieldMappings()
# Creating an individual field map for each field and adding it to the
field mapping object
for field in myfields :
    map = arcpy.FieldMap()
    map.addInputField(inFeatures, field)
    mapS.addFieldMap(map)
    # Copying the feature class using the fields
    arcpy.FeatureClassToFeatureClass_conversion(inFeatures,
    outLocation, outFeatureClass, field_mapping=mapS)
#Joining of Adaptive capacity 2001 field to the Adaptive capacity
2011 field to create one table
arcpy.JoinField_management ("AD_Diff", "MP_NAME",
"Ind_Scores2001", "MP_NAME", "AD2001_Score")
    ## for years 2001 and 2011 into one attribute table
    #Importing system modules
    import arcpy
    from arcpy import env
    #Setting the environment
env.workspace = "E:/GIS/"
    # Specifying of input feature class, output location and feature
    classes
    inFeatures = "E:/GIS/RayMhlaba.mdb/Ind_Scores2011"
    outLocation = "E:/GIS/RayMhlaba.mdb"
    outFeatureClass = "AD_Diff"
    # Listing of fields to be retained
myfields = ["MP_NAME", "AD2011_Score"]
    # Creating an empty field mapping object
mapS = arcpy.FieldMappings()
    # Creating an individual field map for each field and adding it to
    the field mapping object
    for field in myfields:
map = arcpy.FieldMap()
map.addInputField(inFeatures,field)
mapS.addFieldMap(map)
    # Copying the feature class using the fields
arcpy.FeatureClassToFeatureClass_conversion(inFeatures,
outLocation,outFeatureClass,field_mapping=mapS)
    #Joining of Adaptive capacity 2001 field to the Adaptive capacity
    2011 field to create one table
arcpy.JoinField_management
("AD_Diff","MP_NAME","Ind_Scores2001","MP_NAME","AD2001_Score")
```

Changes in adaptive capacity scores between census years 2001 and 2011 (Table 3) were calculated by using the following formula:

$$\text{Change in adaptive capacity} = \left(\begin{array}{c} 2011 \\ \sum_{n=s1}^{s4} n \end{array} \right) - \left(\begin{array}{c} 2001 \\ \sum_{n=s1}^{s4} n \end{array} \right) \tag{2}$$

The adaptive capacity scores were automatically ranked into no change, decrease, and increase as follows: Difference of 0 = NO CHANGE; Difference of ≤ -1 = DECREASE; Difference of ≥ 1 = INCREASE by using the following Python script (Script 5) that was run by executing the Python `execfile` command which also generated fields for differences and ratings of adaptive capacity scores for the two census years and automatically added them to the attribute table.

Script 5 Script That Was Used for Calculating Changes in Adaptive Capacities Between 2001 and 2011

```
## Name of script: Diff.py
## Purpose: Calculating changes in adaptive capacities between
years 2001 to 2011
#Importing system modules
import arcpy, math
# Selecting fields of interest from attribute table in shapefile
table = "E:/GIS/RayMhlaba.mdb/AD_Diff"
fields = ["AD2011_Score","AD2001_Score"]
# Adding new fields to table to store calculated differences in
adaptive capacity
difference = "AC_Resultant"
change = "AC_Change"
arcpy.AddField_management(table, difference, "SHORT")
arcpy.AddField_management(table, change, "TEXT")
#Adding created fields to the array
fields2 = fields[:]
fields2.extend([difference, change])
#Classifying the changes in adaptive capacity
with arcpy.da.UpdateCursor(table, fields2) as cursor:
    for row in cursor:
        arrayVals = [row[0], row[1]]
        #Subtracting the 2001 adaptive capacity from the 2011
        adaptive capacity
    row[2] = row[0] - row[1]
    #Allocating the adaptive capacity change
    if row[2] <= -1:
        row[3] = 'DECREASE'
        cursor.updateRow(row)
    elif row[2] == 0:
        row[3] = 'NO CHANGE'
        cursor.updateRow(row)
    else:
        row[3]='INCREASE'
```

```
        cursor.updateRow(row)
            #Importing system modules
import arcpy,math
            # Selecting fields of interest from attribute table in
            shapefile
            table = "E:/GIS/RayMhlaba.mdb/AD_Diff"
fields = ["AD2011_Score","AD2001_Score"]
            # Adding new fields to table to store calculated
            differences in adaptive capacity
            difference = "AC_Resultant"
            change = "AC_Change"
arcpy.AddField_management(table,difference,"SHORT")
arcpy.AddField_management(table,change,"TEXT")
            #Adding created fields to the array
            fields2 = fields[:]
fields2.extend([difference,change])
            #Classifying the changes in adaptive capacity
with arcpy.da.UpdateCursor(table,fields2) as cursor:
            for row in cursor:
            arrayVals = [row[0],row[1]]
            #Subtracting the 2001 adaptive capacity from the 2011
            adaptive capacity
            row [2] =row[0] -row[1]
            #Allocating the adaptive capacity change
            if row [2] <= - 1:
            row [3] = 'DECREASE'
cursor.updateRow(row)
            elif row [2] ==0:
            row [3] = 'NO CHANGE'
cursor.updateRow(row)
            else:
            row [3] = 'INCREASE'
cursor.updateRow(row)
```

A map showing changes in adaptive capacities was produced in ArcMap 10.5 based on the rankings and another map showing communities with low adaptive capacities by year 2011 produced from the 2011 adaptive capacity attribute table. Thereafter, the same attribute table and platform were used to generate a list of communities with decreased adaptive capacities over the 10 years between 2001 and 2011.

Presentation of Results

Results of this study are presented in the form of (1) maps that were purposefully designed to illustrate how tabulated data can be portrayed in a visually comprehensible manner that facilitates better tracking of adaptation compared to conventional presentation of this important information in the form of tables and histograms that do not capture the spatial dimension (Figs. 2, 3, and 4) and (2) tables that summarize communities that were identified as having declining/static adaptive capacities and

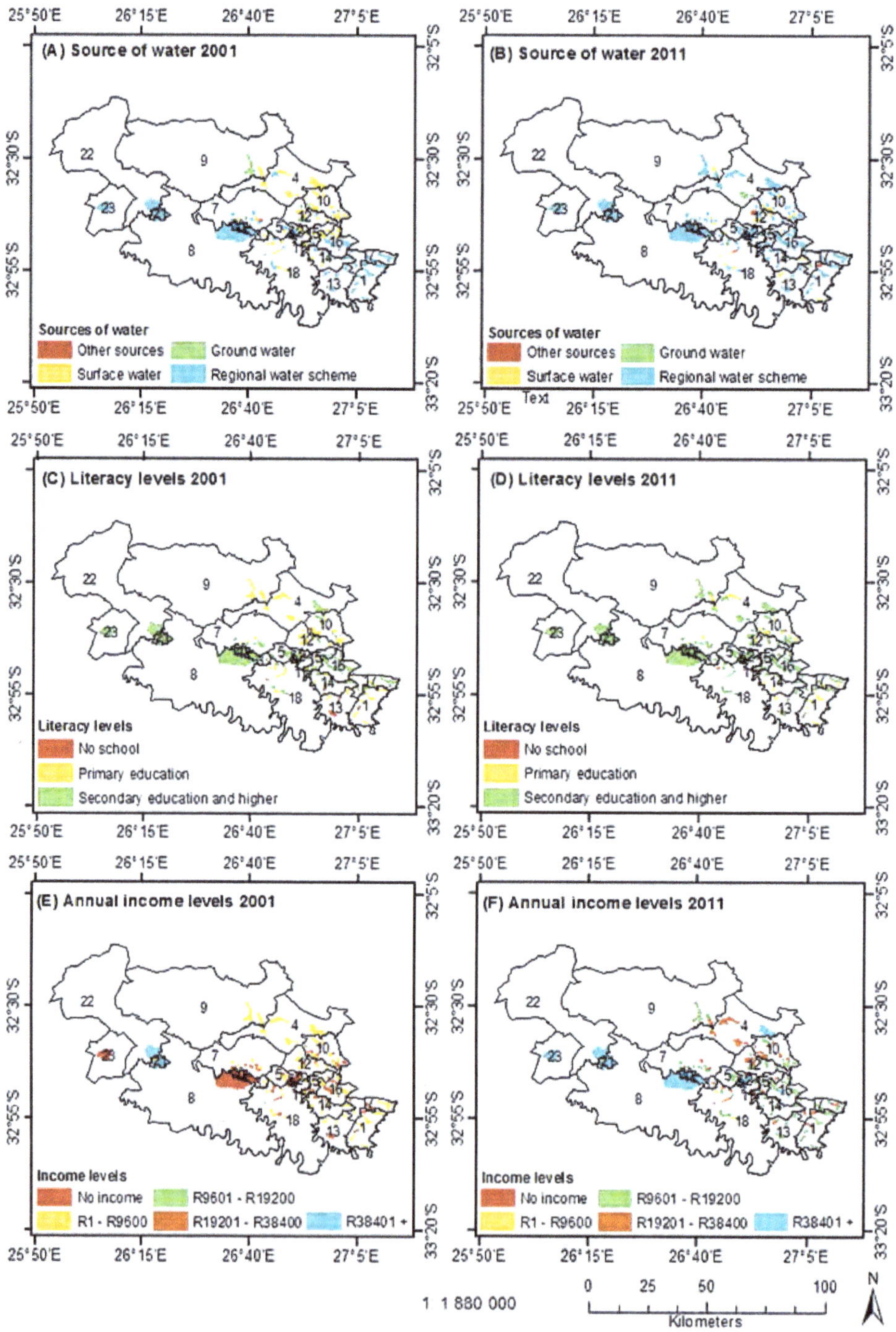

Fig. 2 Access to water by source type, literacy, and annual income levels for communities in RMLM in 2001 and 2011. (Source: Chari 2020)

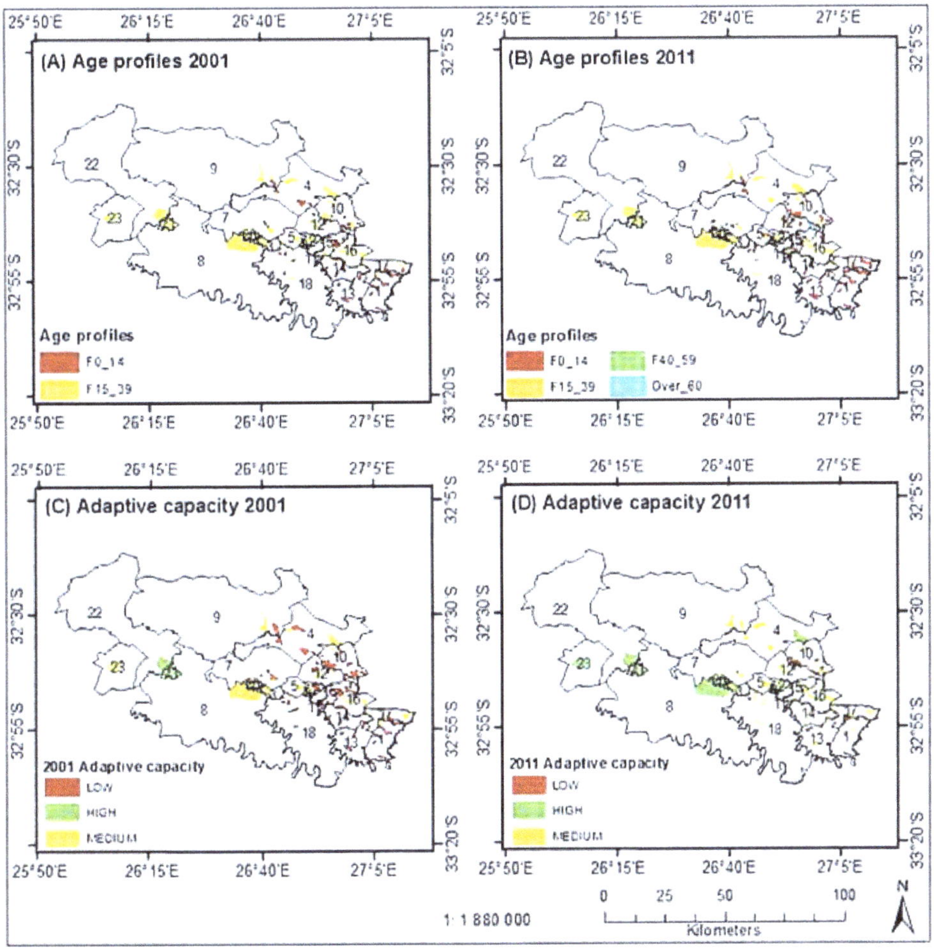

Fig. 3 Age profiles and adaptive capacities of communities in RMLM in 2001 and 2011. (Source: Chari 2020)

those that were identified as having long-term decrease in adaptive capacities (Tables 4 and 5, respectively).

Maps That Were Produced from Data on Water Access, Literacy Levels, and Annual Income Levels That Were Used to Assess Adaptive Capacity for Years 2001 and 2011

Figure 2 shows spatial distributions of access to water by source type, literacy levels, and annual incomes that were mapped for individual communities in RMLM on the basis of census data for the years 2001 and 2011. Figure 3 shows age profiles for the same communities and their adaptive capacities on the basis of the three-point scale (LOW-MEDIUM-HIGH) that was used to rank the uptake of adaptation interventions by individual communities. The last map (Fig. 4) shows spatial variations in

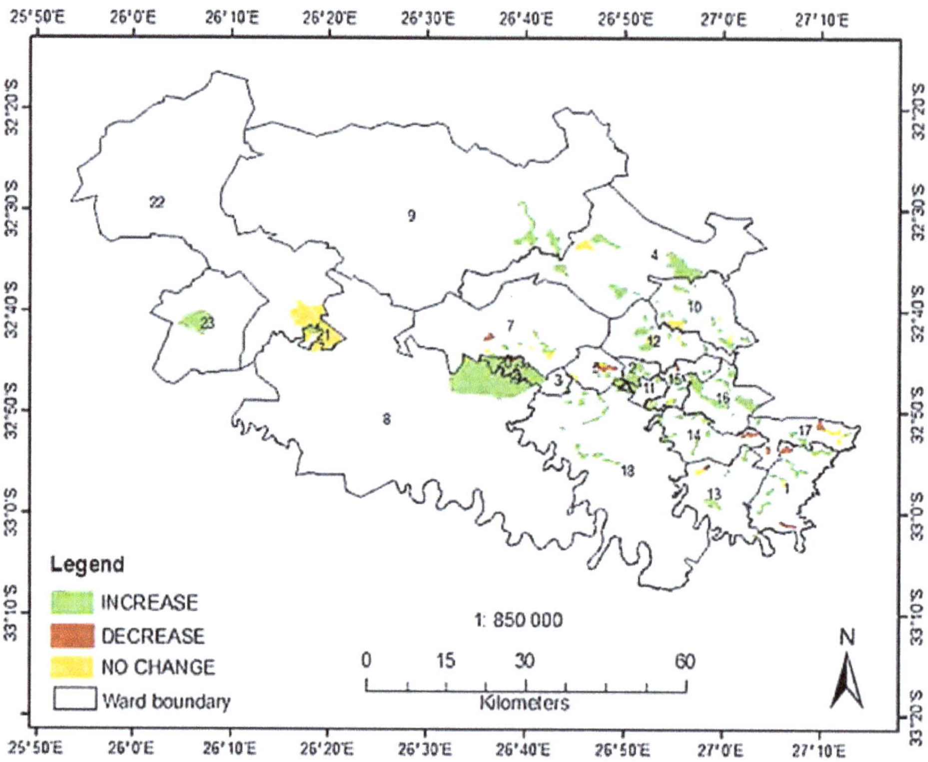

Fig. 4 Changes in adaptive capacities between years 2001 and 2011. (Source: Chari 2020)

Table 4 Communities that were identified as having a long-term decrease in adaptive capacities from 2001 to 2011

Name of community by number of ward in which it is located	Ward number
MnqabaJames	1
Qamdobowa	1
Mgquba	5
Fernvilla	7
eMgwanisheni	13
Newtown	14
Ncera	15
Koloni	17
Zihlahleni	17
KwaMlalandle	20
N = 10	

Source: Chari 2020

abilities of individual communities to mitigate the adverse effects of climate change by capturing spatial distributions of changes in adaptive capacities that were identified from the ranked adaptive capacity scores in Table 3.

Table 5 Communities that were identified as having declining/static adaptive capacities in Raymond Mhlaba Local Municipality in 2001 and 2011

Name of community by number of ward in which it is located			Adaptive capacities 2001	Adaptive capacities 2011
KwaKulile	1		LOW	LOW
Mnqaba James	1		MEDIUM	LOW
Qamdobowa	1		LOW	LOW
KwaNacelwane	7		LOW	LOW
Jomlo	10		LOW	LOW
Lower Hopefield	10		LOW	LOW
Machibini	10		LOW	LOW
Mdeni B	10		LOW	LOW
Mazotshweni	12		LOW	LOW
eMgwanisheni	13		MEDIUM	LOW
KuDikidikana	13		LOW	LOW
Lebanon	18		LOW	LOW
N = 12				

Source: Chari 2020

Tables That Show Adaptive Capacity Levels for Individual Communities by Year Period

Tables 4 and 5 show communities that were identified as having (a) declining (medium – low) and static (low) adaptive capacities in the 2001 and 2011 census surveys and (b) communities that were identified as having a long-term decrease in adaptive capacity in the 10 years between 2001 and 2011.

Discussion and Conclusion

Discussion

This section discusses the results of this initiative by placing the major findings in a broader context with emphasis being given on the responsiveness of communities to climate change on the basis of the extent to which individual communities were able to access the three indicators that were singled out for investigation and how age profiles influence their adaptive capacities. The section concludes the chapter by stating the methodology's limitations and highlighting its usefulness in tracking adaptation.

Access to Water

Sources of water by source type in wards 1–23 (Fig. 2a and b) affect the resilience of communities (not shown in Fig. 2a and b) by influencing the availability of water. In 2001 (Fig. 2a), two communities (Teba and KwaNgwevu in ward 7) were severely water stressed and obtained water from two nonnatural sources comprising water vendors and water tankers. By 2011 (Fig. 2b), three communities (Msobomvu in ward 12, Allandale in ward 13, and MnqabaJames in ward 1) were identified as severely water stressed, although they still depended on water from the same nonnatural sources as they did in 2001. The increase in the number of communities depending on nonnatural sources from 2 to 3 is indicative of how deteriorating climatic conditions are reducing natural availability of water which is corroborated by the observed occurrence of consecutive droughts in the Eastern Cape province in 1992, 2004, and 2009 (ADM 2010, 2012; IFRC 2004). These scenarios strongly suggest that in this environment, water scarcity is a persistent problem that requires exploration of alternative water sources and a judicious mix of different water saving techniques in order to reduce dependence on costly supplies provided by government and commercial operators.

Literacy Levels

Figure 2c and 2d indicates spatial variations in levels of schooling. In 2001 (Fig. 2c), two communities (NgqolowaA and Dhlawu) in ward 13 and six communities (Ndindwa, Gqumashe, KwaMemela, Lower Hopefield, Mavuvumezini, and Calderwood) in wards 1, 2, 3, 10, 14, and 18, respectively, were identified as having no formal schooling. By 2011, only two communities (Mdeni B in ward 5 and Lebanon in ward 13) were identified as having the majority of the population without formal schooling (Fig. 2d). This observation is not only commendable but interesting because it suggests that awareness of climate change issues is increasing as the number of people gaining access to formal education increases. Unfortunately, awareness alone does not automatically translate into increased uptake of actionable adaptation strategies in the absence of enabling factors that facilitate the translation of recommended actions into tangible climate friendly activities because knowing alone without inspiration and capacity to act is inadequate (Adger et al. 2009). Ability to act requires capacity in the form of resources which in the case of RMLM is evidently lacking because of poverty. This asseveration is supported by the 2011 census data which shows that 48.1% of the people in Nkonkobe and 42% in Nxuba were unemployed (StatsSA 2014). In view of this consideration and the documented prevalence of poverty in this area, it is not unreasonable to suggest that although literacy improved, effective implementation of climate change adaptation strategies continues to be constrained by lack of a holistic approach that embraces a wide range of enabling factors that determine the ability of communities to meaningfully embrace climate-friendly interventions.

Annual Income Levels

Figure 2e and 2f indicate communities with and without income, respectively, with the latter being destined to be susceptible to most climate change-induced shocks due to lack of access to credit. In year 2001, 51 communities emerged as being dominated by people without any income (Fig. 2e) while 5 communities (Seymour, MdeniA, Magxagxeni, Koloni, and kwaSawu) in wards 4, 7, 10, 17, and 15, respectively, were identified as the poorest in the municipality (Fig. 2f). These results are in agreement with the 2011 census data which shows that the majority of people in these communities do not have any sources of income (StatsSA 2014). The nationwide prevalence of this limitation is supported by many researchers (Mkuhlani et al. 2019; Rusere et al. 2019; Mpandeli and Maponya 2013) who report that in the Lambani, Tshakhuma, Rabali, and Tshiombo communities in Limpopo province, most households find rain-fed crop production extremely risky because they cannot afford supplementary irrigation and do not qualify to get external support in the form of agricultural insurance due to widespread lack of reliable sources of income.

Influence of Age Profiles on Resilience

The identification of communities with different levels of resilience on the basis of age profiles (Fig. 3a and 3b) was based on the reasoning that children and old people have limited capacities to assimilate and implement adaptation strategies by virtue of being economically inactive compared to their economically active counterparts of intermediate age. In year 2001 (Fig. 3a), 37 communities were identified as having the majority of the people in the ages between 0 and 14 years. By year 2011 (Fig. 3b), a total of 46 communities were identified as having the majority of people in the ages between 0 and 14 years with 2 of these communities having the majority of people above 60 years of age while 68 communities had low adaptive capacities in 2001 (Fig. 3c). Although 54 of these communities had improved from low to medium and high adaptive capacities by 2011, this apparent improvement falls short of the desired situation because the downward trend in capacities of two of these communities (eMgwanisheni and MnqabaJames) is actually indicative of the system's inability to effectively mitigate the adverse effects of climate change with additional support for this observation coming from the fact that between 2001 and 2011, ~7.5% of the communities in RMLM were identified as having low or decreasing adaptive capacities (Table 5).

The results presented here are useful because they demonstrate that objectively based geostatistical techniques can be used to aid disaster management by creating space for timely and reliable identification of communities that can be targeted as recipients of appropriately informed adaptation strategies at multiple temporal and spatial scales. In a broader context, the methodology is potentially capable of enhancing adaptation tracking because of its ability to accommodate wide-ranging demographic datasets that are often provided in different and unstandardized formats. This proficiency makes it adaptable to suit different areas of interest and capable of being used to support different stakeholders such as the Global Commission on Adaptation and many others which seek to ensure that adaptation action and support reaches the most vulnerable communities.

Conclusion

Although the methodology presented in this chapter is admittedly far from being capable of offering a universal solution due to gaps in the availability of data at appropriate spatial and temporal scales and lack of a single unit of analysis that can be used to measure or compare adaptation, accessibility of demographic data in most countries still makes it a usable option worth considering because the acquisition of fine-scale data can be prohibitively expensive and time consuming. Even though these challenges are likely to persist for some time into the future, one of the major strengths of the methodology is that it provides an ideal entry point to adaptation tracking at both the national and sub-national levels because its bottom-up basis and firm grounding in a GIS-based framework can accommodate consistent and contextually focused multilevel assessment at different spatial and temporal scales. The other strength of the methodology is that its bottom-up inclination confers a sense of involvement in the decision-making process by creating space for the incorporation of inputs from key stakeholders who are directly involved in and affected by the implementation of adaptation projects. In addition, the methodology's ability to provide a spatially explicit presentation of adaptive capacities offers additional advantages by enabling those interested in adaptation tracking to identify trends in the assimilation and implementation of climate friendly adaptation interventions. The major insight from this initiative is that although data scarcity is often singled out as one of the major constraints confronting adaptation tracking, the geostatistical methodology presented in this chapter provides a workaround approach that can be tapped and used to maximize the utility of different and readily accessible datasets from disparate sources. The take home message is that practitioners, policy–makers, and other stakeholders interested in fast-tracking and enhancing the effectiveness of adaptation interventions stand to benefit by adopting this methodology because it can be easily adapted to effectively track the adoption of adaptation under wide-ranging situations.

Acknowledgments The authors would like to thank the Water Research Commission's K5/2496/4 project and National Research Foundation (NRF) of South Africa for funding.

References

Adger WN, Dessai S, Goulden M, Hulme M, Lorenzoni I, Nelson DR, Naess LO, Wolf J, Wreford A (2009) Are there social limits to adaptation to climate change? Clim Chang 93(3):335–354

ADM (2010) Amathole District Municipality State of the District Address. Amathole Community News, pp 1–2. Available http://www.amathole.gov.za/old/attachments/article/70/English Newsletter 2010 June.pdf. Accessed 21 Jan 2019

ADM (2012) Amathole District Municipality 2012–2017 Integrated Development Plan. Available http://www.amathole.gov.za/old/attachments/article/324/FINAL%20IDP%202012-2017.pdf. Accessed 04 Feb 2019

ADM (2017) Amathole District Municipality 2017–2020 Integrated Development Plan (IDP) – Eastern Cape Socio Economic Consultative Council (ECSECC). Available https://www.ecsecc.org/information-centre/item/amathole-2017-2020-integrated-development-plan-idp. Accessed 28 Jan 2019

Berrang-Ford L, James D, Ford JD, Paterson J (2011) Are we adapting to climate change? Glob Environ Chang 21(1):25–33

Berrang-Ford L, Biesbroek, Ford JD, Lesnikowski A, Tanabe A, Wang FM, Chen C, Hsu A, Hellmann JJ, Pringle P, Grecequet M (2019) Tracking global climate change adaptation among governments. Nat Clim Chang 9:440–449

Biesbroek G, Swart RJ, Carter TR, Cowan C, Henrichs T, Mela H, Morecroft MD, Rey D (2010) Europe adapts to climate change: comparing national adaptation strategies. Glob Environ Chang 20(3):440–450

Bouroncle C, Imbach P, Rodríguez-Sánchez B, Medellín C, Martinez-Valle A, Läderach P (2017) Mapping climate change adaptive capacity and vulnerability of smallholder agricultural livelihoods in Central America: ranking and descriptive approaches to support adaptation strategies. Clim Chang 141(1):123–137

Brooks N (2011) Tracking adaptation and measuring development. IIED climate change working paper: no. 1. IIED, London

Chari MM (2020) Enhancing adaptive capacities of farmers to climate-induced rainfall variabilities by modelling soil moisture patterns in Raymond Mhlaba Local Municipality, Eastern Cape Province, South Africa. Draft PhD thesis, Department of Geography & Environmental Science, University of Fort Hare, Alice main campus, South Africa

Chari MM, Hamandawana H, Zhou L (2018) Using geostatistical techniques to map adaptive capacities of resource-poor communities to climate change: a case study of Nkonkobe local municipality, Eastern Cape Province, South Africa. Int J Clim Change Strategies Manage 10 (5):670–688

Conway D, Nicholls RJ, Brown S, Tebboth MG, Adger WN, Ahmad B, Biemans H, Crick F, Lutz AF, De Campos RS, Said M (2019) The need for bottom-up assessments of climate risks and adaptation in climate-sensitive regions. Nat Clim Chang 9:503–511

de Sherbinin A (2016) Remote sensing and socioeconomic data integration: lessons from the NASA socioeconomic data and applications center. CRC Press, Boca Raton

Dinar A, Hassan R, Mendelsohn R, Benhin J (2012) Climate change and agriculture in Africa: impact assessment and adaptation strategies. Routledge. Available https://content.taylorfrancis.com/books/download?dac=C2011-0-06085-0&isbn=9781136569746&format=googlePreviewPdf. Accessed 10 Apr 2019

Dwesini X (2018) ADM secures 10 trucks for drought relief. Daily Dispatch, South Africa. https://www.dispatchlive.co.za/news/2018-09-18-adm-secures-10-trucks-for-drought-relief/. Accessed 04 Oct 2019

Engelbrecht F, Adegoke J, Bopape MJ, Naidoo M, Garland R, Thatcher M, McGregor J, Katzfey J, Werner M, Ichoku C, Gatebe C (2015) Projections of rapidly rising surface temperatures over Africa under low mitigation. Environ Res Lett 10(8):85004

FAO (2017) Tracking adaptation in agricultural sectors: Climate change adaptation indicators. Food and Agriculture Organization of the United Nations (FAO), Rome. Available http://www.fao.org/resilience/resources/resources-detail/en/c/1106739/. Accessed 03 April 2019

Ford J, Berrang-Ford L, Lesnikowski A, Barrera M, Heymann S (2013) How to track adaptation to climate change: a typology of approaches for national-level application. Ecol Soc 18(3):40

Ford JD, Berrang-Ford L, Bunce A, McKay C, Irwin M, Pearce T (2015) The status of climate change adaptation in Africa and Asia. Reg Environ Chang 15(5):801–814

Fox KM, Nelson S, Frankenberger TR, Langworthy M (2018) Climate change adaptation in Ethiopia: developing a method to assess program options. In: Resilience, pp 253–265. https://doi.org/10.1016/B978-0-12-811891-7.00021-9. Elsevier

Gamble JL, Hurley BJ, Schultz PA, Jaglom WS, Krishnan N, Harris M (2013) Climate change and older Americans: state of the science. Environ Health Perspect 121(1):15–22

Gbetibouo GA, Ringler C, Hassan R (2010) Vulnerability of the South African farming sector to climate change and variability: an indicator approach. Nat Res Forum 34(3):175–187

Georgeson L, Maslin M, Poessinouw M, Howard S (2016) Adaptation responses to climate change differ between global megacities. Nat Clim Chang 6(6):584–588

Hamandawana H, Chanda R, Eckardt F (2008) Reappraisal of contemporary perspectives on climate change in southern Africa's Okavango Delta sub-region. J Arid Environ 72(9):1709–1720

Heltberg R, Bonch-Osmolovskiy M (2011) Mapping vulnerability to climate change. World Bank Policy Research, Washington, DC

Hlahla S, Nel A, Hill TR (2018) Assessing municipal-level governance responses to climate change in KwaZulu-Natal, South Africa. J Environ Plan Manag 62(6):1089–1107

https://www.greenclimate.fund/countries/south-africa. Accessed 17 Apr 2019

IFRC (International Federation of Red Cross) (2004) South Africa: drought. Available http://www.ifrc.org/docs/appeals/rpts04/ZA040130.pdf. Accessed 03 Jan 2019

IPCC (2014) Climate change 2014: synthesis report. Contribution of Working Groups I, II and III to the fifth assessment report of the Intergovernmental Panel on Climate Change [Core writing team, Pachauri RK, Meyer LA (eds)]. Geneva. Available https://www.ipcc.ch/report/ar5/syr/. Accessed 13 Apr 2019

IPCC (2018) Summary for policymakers. In: Masson-Delmotte V, Zhai P, Pörtner H-O, Roberts D, Skea J, Shukla PR, Pirani A, Moufouma-Okia W, Péan C, Pidcock R, Connors S, Matthews JBR, Chen Y, Zhou X, Gomis MI, Lonnoy E, Maycock MT, Waterfield T (eds) Global warming of 1.5°C. An IPCC Special report on the impacts of global warming of 1.5°C above pre-industrial levels and related global greenhouse gas emission pathways, in the context of strengthening the global response to the threat of climate change, sustainable development, and efforts to eradicate poverty. World Meteorological Organization, Geneva, 32 pp. Available https://www.ipcc.ch/sr15/chapter/summary-for-policy-makers/. Accessed 23 Nov 2018

Jiri O, Mafongoya P (2017) A synthesis of smallholder farmers' adaptation to climate change in Southern Africa: averting adaptation vacuum. In: Climate change adaptation in Africa. Springer, Cham, pp 247–265. https://doi.org/10.1007/978-3-319-49520-0_16

Juhola S, Kruse S (2015) A framework for analysing regional adaptive capacity assessments: challenges for methodology and policy making. Mitig Adapt Strateg Glob Chang 20(1):99–120

Jury MR (2019) South Africa's future climate: trends and projections. In: Knight J, Rogerson CM (eds) The geography of South Africa. Contemporary changes and new directions. World regional geography book series, vol 4. Springer International Publishing AG, a part of Springer Nature, Cham, pp 305–312

Keskitalo E, Preston B (2019) Research handbook on climate change adaptation policy. Edward Elgar Publishing, Cheltenham/Northampton

Kuwornu JKM (2019) Climate change and Sub-Saharan Africa: the vulnerability and adaptation of food supply chain actors series on climate change and society. Vernon Press, Wilmington

Leal Filho W, de Freitas LE (2018) Climate change adaptation in Latin America. Springer, Cham

Leal Filho W, Esilaba AO, Rao KP, Sridhar G (2015) Adapting African agriculture to climate change. Springer, Berlin, p 233

Leal Filho W, Simane B, Kalangu J, Wuta M, Munishi P, Musiyiwa K (2017) Climate change adaptation in Africa. Climate change management, 1610–2010. Springer, Cham

Lesnikowski A, Ford J, Biesbroek R, Berrang-Ford L, Maillet M, Araos M, Austin SE (2017) What does the Paris Agreement mean for adaptation? Clim Pol 17(7):825–831

Local Government Handbook (2016) The local government handbook: South Africa. http://www.localgovernment.co.za/locals/view/233/Raymond-Mhlaba-Local-Municipality. Accessed 27 Feb 2019

Mkuhlani S, Crespo O, Rusere F, Zhou L, Francis J (2019) Classification of small-scale farmers for improved rainfall variability management in South Africa. Agroeco Sustain Food Syst 1–23. https://doi.org/10.1080/21683565.2018.1537325

Mpandeli NS, Maponya PI (2013) Coping with climate variability in Limpopo Province, South Africa. Peak J Agric Sci 1:54–64

Mukarakate D (2016) Global climate finance – will Africa benefit after Paris?. http://www.africa.undp.org/content/rba/en/home/blog/2016/4/21/Global-Climate-Finance-Will-Africa-Benefit-After-Paris-.html. Accessed 25 Mar 2019

Ncube M, Madubula N, Ngwenya H, Zinyengere N, Zhou L, Francis J, Mthunzi T, Olivier C, Madzivhandila T (2016) Climate change, household vulnerability and smart agriculture: the case of two South African provinces. Jàmbá J Disaster Risk Stud 8(2):6–8

Ndamase M (2019) ADM again declared a drought disaster area. Daily Dispatch, South Africa. https://www.pressreader.com/south-africa/daily-dispatch/20190710/281522227653276. Accessed 10 Oct 2019

New M (2018) What the latest assessment on global warming means for southern Africa. Quest 14 (4):12–13

Ngqakamba S (2019) Gift of Givers aids drought-stricken Amathole municipality with bottled water in truckloads. News 24, South Africa. https://www.news24.com/SouthAfrica/News/gift-of-givers-aids-drought-stricken-amathole-municipality-with-bottled-water-in-truckloads-20190410. Accessed 06 Oct 2019

Nkonkobe Local Municipality (2012) Integrated development plan 2012–2017. http://nkonkobe.gov.za/?q=system/files/filedepot/2/FINAL IDP 2012–2017.pdf. Accessed 23 Jan 2019

Pauw WP, Klein RJ, Mbeva K, Dzebo A, Cassanmagnago D, Rudloff A (2018) Beyond headline mitigation numbers: we need more transparent and comparable NDCs to achieve the Paris Agreement on climate change. Clim Chang 147:23–29

Qiu Y, Zhao X, Fan D, Li S (2019) Geospatial disaggregation of population data in supporting SDG assessments: a case study from Deqing County, China. ISPRS Int J Geo Inf 8(8):356

Rankoana SA (2019) Food security under unreliable rainfall: the case study of a rural community in Limpopo Province, South Africa. J Water Clim Change. https://doi.org/10.2166/wcc.2019.109

Rusere F, Mkuhlani S, Crespo O, Dicks LV (2019) Developing pathways to improve smallholder agricultural productivity through ecological intensification technologies in semi-arid Limpopo, South Africa. Afr J Sci Technol Innov Dev 1–11. https://doi.org/10.1080/20421338.2018.1550936

Sarkodie SA, Strezov V (2019) Economic, social and governance adaptation readiness for mitigation of climate change vulnerability: evidence from 192 countries. Sci Total Environ 656:150–164

Schensul D, Martine G, Guzmán JM (2013) The demography of adaptation to climate change. UNFPA, IIED, and El Colegio de México, New York

Statistics South Africa (2014) Census 2011 provincial profile: Eastern Cape. Report No. 03-01-71. http://www.statssa.gov.za/publications/Report-03-01-71/Report-03-01-712011.pdf. Accessed 20 July 2019

Taylor A (2016) Institutional inertia in a changing climate. Int J Clim Change Strategies Manage 8 (2):194–211

Tompkins EL, Vincent K, Nicholls RJ, Suckall N (2018) Documenting the state of adaptation for the global stocktake of the Paris Agreement. Wiley Interdiscip Rev Clim Chang 9(5):e545

Van Aalst MK, Cannon T, Burton I (2008) Community level adaptation to climate change: the potential role of participatory community risk assessment. Glob Environ Chang 18(1):165–179

Wall E, Marzall K (2006) Adaptive capacity for climate change in Canadian rural communities. Local Environ 11(4):373–397

Weis SW, Agostini VN, Roth LM, Gilmer B, Schill SR, Knowles JE, Blyther R (2016) Assessing vulnerability: an integrated approach for mapping adaptive capacity, sensitivity, and exposure. Clim Chang 136(3):615–629

World Bank (2014) Turn down the heat: confronting the new climate normal. World Bank Publications, Washington, DC. www.statssa.gov.za. Accessed 13 Mar2019

Zhou L, Sibanda M, Musemwa L, Ndhleve S (2016) Vulnerability to climate change related disasters in the Eastern Cape Province: an application of the Household Vulnerability Index (HVI). J Hum Ecol 56(3):335–353

Ziervogel G, New M, Archer van Garderen E, Midgley G, Taylor A, Hamann R, Stuart-Hill S, Myers J, Warburton M (2014) Climate change impacts and adaptation in South Africa. Wiley Interdiscip Rev Clim Chang 5(5):605–620

3

Uncertainties in Rainfall and Water Resources in Maghreb Countries Under Climate Change

Mohamed Meddi and Saeid Eslamian

Contents

Abstract

The vulnerability of the climate change in the South of the Mediterranean's south regions varies depending on the part of their climate which is sensitive to the economy. In Tunisia, agriculture represents 16% of the workforce and 12% of GDP in 2006. In Algeria, agriculture represents 20% of the workforce and 8% of GDP in 2009. In Morocco, agriculture accounts for 40% of the workforce and 17.7% of GDP in 2006. The agriculture is directly related to the availability of water which in turn is directly related to rainfall. The drought has affected all countries of the Maghreb. It is considered the most severe in the history of these

M. Meddi (✉)
Ecole Nationale Supérieure d'Hydraulique de BLIDA, Boufarik, Algeria
e-mail: m.meddi@ensh.dz

S. Eslamian
Department of Water Engineering, College of Agriculture, Center of Excellence on Risk Management and Natural Hazards, Isfahan University of Technology, Isfahan, Iran
e-mail: saeid@cc.iut.ac.ir

countries. The drought has forced the agricultural sector in Morocco to the limitation of annual crops which are not needed, the prohibition of any new tree planting and the ban on vegetable crops in dry years. During the years 1987, 1988, and 1989, Tunisia has experienced the most critical drought. It led to a water deficit of around 30%. For Morocco the rainfall shows a negative trend at national and regional scales, and spring rainfall has declined by over 40% since the 1960s. For Algeria, the western region has recorded a considerable reduction in rainfall. The winter rains have decreased between 40% and 70%. Contributions to dams have decreased between 30% and 50%. These changes had a negative influence on the water resource and crop yield. Many programs have been initiated since then to meet the growing demand.

Keywords

Rainfall · Water resources · Drought · Climate change · Maghreb countries

Introduction

The issue of climate change settled permanently at the forefront of the news and concerns opinion's public. Confined for a long time to the scientific community and environmental groups, it has emerged in recent years at a government level with the signing of the Kyoto Protocol. But it is mostly the recent weather disturbances (repeated heat waves, devastating hurricanes). Meanwhile, advances in scientific knowledge have gradually transformed what was originally a simple hypothesis – a global warming caused by man – a virtual certainty, although there are still many unknowns about the accurate scope and the pace of the phenomenon.

According to the World Bank (2017), Tunisia, Algeria, and Morocco will become global hot spots by the end of the twenty-first century. Higher temperatures and reduced precipitation will increase the occurrence of droughts in this region (World Bank 2017).

Droogers et al. (2012) found that climatic changes will be responsible for 22% of water shortages by 2050 and 78% of water shortages will be the result of socioeconomic factors for North Africa. During the last century, Algeria has experienced several periods of drought, the most intense were felt in 1910 and 1940 and more persistent in the years from 1975 to 1990 and the beginning of the century. The last drought in the Maghreb countries since five past decades is particularly severe. The period 1950–1975 was wetter comparing to the long-term average, so that from 1977 there is an enough marked drought, unprecedented since the early observations. Algeria and especially the West region have experienced several major droughts in the last century, during the 1940s and 1970s until today (Meddi and Hubert 2003; Habibi et al. 2018; Merabti et al. 2018). The latest one was characterized by its spatial extent and its intensity. In Morocco, droughts continue to increase in both frequency and intensity and impact negatively water resources and agricultural production according to World Bank (2017). Droughts are recurrent and are becoming more frequent and more severe. Previous recorded droughts, which required government emergency plans, were recorded in 1992–1995, 1998–2001, 2005, 2007, and 2015–2016. The 1994–1995 droughts were particularly strong, resulting in an estimated GDP loss of 7.6% (World Bank Group 2017a). The 1999 drought cost about US$ 900 million and affected more than one million hectares of cultivated land (World Bank Group 2017a).

To the large spatial variability of precipitations correspond a clear regional difference in the potential of surface water. At the seasonal and interannual variability of rainfall, there is also a strong irregular hydrographical inflow with violent floods and low water periods that can last several months (Meddi and Hubert 2003).

Two-thirds of the territory of Tunisia is characterized by a semiarid climate to arid. This space is subject to drought may be common. The droughts are a common natural phenomenon in Tunisia. They become more and more frequent. The resulting consequences have increasingly serious social and economic impacts. The droughts date back to the last century and Tunisia experienced 31 dry years between 1907 and 2016. More recently, Tunisia experienced a drought during the years: 1982, 1987 to 1989, 1993, 1994, 1995, 1997, 2000, 2001, 2002, 2008, 2010, 2013, and 2016, and the worst drought since more than 50 years was recorded during the years from 1999 to 2002. These droughts cost about $ 54 million.

The agricultural sector is largely dependent on the availability of water resources, which is the most vulnerable to climate change (Farzaneh et al. 2014). In Algeria, the water consumption in the agricultural sector is about 7 billion m^3/year where it represents 70% of the total national consumption (http://www.mree.gov.dz). In Morocco, irrigation uses 80% of the country's mobilized resources, the rest being reserved for drinking water and industry (MDCE 2016). The used water in irrigation is about 12,038 million m^3 (Dahan 2017). In Tunisia, the volume of water intended for irrigation id about 80% of total water consumption (Chebil et al. 2019). These volumes of water used in irrigation to ensure food security and socioeconomic development in these countries demonstrate the importance of studies and the proposal of scenarios on the impact of climate change on water resources in order to be able to propose measures meet the growing demand in the medium and long term.

Climate projections confirm reduction of rainfall and rising of temperatures (Matari 2016), which will affect the availability of water resources in the medium and long term, hence the need to find ways to mobilize additional water resources (conventional and unconventional) to meet this ever increasing demand.

In this chapter, we will discuss the effects of climate change on water resources as well as adaptation measures in the three Maghreb countries.

Maghreb Countries, Water Resources, and Climate Change

The three Maghreb countries (Algeria, Tunisia, and Morocco) occupy an area of 3,254,000 km^2 (Fig. 1). The Maghreb is crossed from north to west by a mountain system called "Atlas." It is a barrier between the Mediterranean coast and the Sahara. It begins at the mouth of the Wadi Sous in southwestern Morocco to Cape Bon and

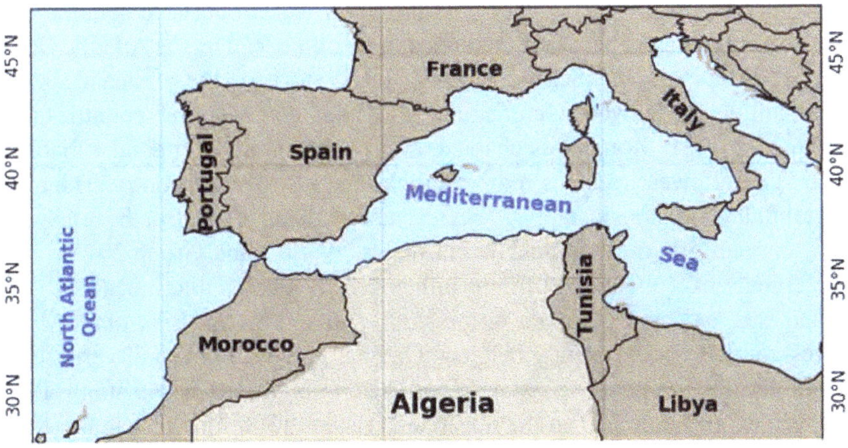

Fig. 1 Situation of Algeria, Tunisia, and Morocco (Zeroual et al. 2017)

the Gulf of Gabes in northeastern Tunisia. South of the Tell, there is the Saharan Atlas which separates the Algerian highlands from the Sahara. It continues its extension to the east by the Aurès and the Tunisian Dorsal. The Sahara occupies nearly 80 percent of the total area. The most fertile agricultural land is located on the coastal strip that separates the mountains from the sea and the highlands.

These three countries are characterized by considerable demographic growth. They are also facing the phenomenon of climate change which has led to a decrease in rainfall and an increase in temperatures. Rainfall in the Maghreb is already low, with average rainfall volumes of about 357 km^3/year, or 110 mm (Taabni and El Jihad 2015).

Tunisia is located between longitudes 7° and 12° east and 32° and 38° latitudes north. Humid to subhumid areas are in the north. The subhumid to semiarid regions are located in the northwest and the Cape Bon region. The semiarid to arid part is located in central Tunisia. The desert covers the south of Tunisia. In Tunisia, rainfall and temperatures vary considerably from north to south (humid coastal zone to desert conditions in the south) and from east to west. They are influenced by the Mediterranean Sea in the north and in the east and by the Sahara in the south and in the southwest. The rains fall mainly between the months of October and May. Temperatures in the arid and semiarid regions of the south and southwest of the country are generally high, while precipitation is much higher from November to April in the more northerly regions.

The northern part of Algeria is characterized by a Mediterranean climate with a relatively cold and rainy winter and a hot and dry summer. The annual rainfall reaches 400 mm in the west, 700 mm in the center, and 1000 mm in the east for the coast. This type of climate is also found in the Tellian Atlas chains where we record totals ranging from 800 to 1600 mm in the eastern summits, while the values are lowered in the center (700 to 1000 mm) and in the west (600 mm). In the plains of the Tellian Atlas, rainfall varies between 500 mm in the west, 450 mm in the center, and 700 mm in the east (Meddi and Hubert 2003).

In Morocco, rainfall follows a decreasing gradient from north to south and from west to east. Rainfall and temperatures are strongly influenced by the Atlantic Ocean to the west, the Mediterranean Sea to the north and the Sahara to the south and southeast. Most rainfall occurs between October and May like other Maghreb countries (World Bank Group 2017a). Morocco is characterized by spatial and temporal variability of rainfall. The northwest receives more rain than the rest of the country. The average annual rainfall varies considerably. It can reach more than 800 mm on the reliefs, while it hardly exceeds 300 mm on the adjacent plains (World Bank Group 2017).

The climate of Tunisia is strongly influenced by the Mediterranean Sea in the north and east and by the Sahara in the south and southwest. Like other Maghreb countries, rainfall occurs mainly between October and May. Rainfall varies considerably from a relatively wet coastal area to desert conditions in the south. Precipitation exceeds 400 mm/year in the north and reaches 1500 mm/year in the extreme northwest of Tunisia (TSNCUNFCCC 2013). In the central part of the country, precipitation varies between 150 and 300 mm/year (TSNCUNFCCC 2013). Rainfall is becoming scarce and less than 150 mm/year and at the extreme south, rains not exceed 50 mm/year. The average total annual rainfall varies from 1 year to another and is around 36 billion m^3/year (TSNCUNFCCC 2013).

The decrease in rainfall since mid-1970s has negatively impacted the renewal of groundwater and the dams filling. The scarcity of water resources has had a dramatic effect on the economy in general and agricultural production and agricultural development in these countries in particular. Agriculture is largely dependent on rainfall. In Morocco, irrigated agriculture contributes about 45% of agricultural added value, despite, irrigated agriculture, occupying only 15% of cultivated area. In Algeria, according to the Ministry of Water Resources, the water consumption in the agricultural sector is close to 7 billion m^3/year where it represents 70% of the total national consumption which is of the order of 10.6 billion m^3/year (http://www.mree.gov.dz). The rate for agriculture corresponds to the world water consumption in irrigation which is around 70%. In Morocco, irrigation uses 80% of the country's mobilized resources, mainly surface water, the rest being reserved for drinking water and industry (MDCE 2016). The used water in irrigation is about 12,038 million m^3 (Dahan 2017). In Tunisia, the volume of water intended for irrigation about 80% of total water consumption (Chebil et al. 2019). This volume was in 2004 of about 2130 million m^3 for an area of 375,000 ha and will be 2145 million m^3 in 2011 and 2035 million m^3 in 2030 for an area of 467,000 ha (www.environnement.gov.tn). In the future, rational pricing encourages farmers to use more and more water-saving technologies and varieties that consume less water.

The renewable water resources in the three Maghreb countries are estimated at 43 km^3, where surface water accounts for 73% (Taabni and El Jihad 2015). In 2010, the average per capita renewable water provision for the Maghreb countries was 548 m^3/person/year in 2010, and by 2030 it will be less than 290 m^3/person/year (Taabni and El Jihad 2015). These values are well below the standard set by the World Health Organization (WHO) of 1000 m^3/person/year.

All these statistics show the importance of the development of water resources management scenarios in the Maghreb countries to meet the increase of water demand for drinking water supply and especially for irrigation, so to increase the capacity of agricultural productivities, and sustainable production systems to reduce the problem of poverty and unemployment knowing that the agricultural sector is the main employer in the world with 40% of the world population.

The Maghreb countries are located in an arid to semiarid region. The climate is Saharian in the southern, oceanic in the western, and Mediterranean in the north. The population of Morocco, Algeria, and Tunisia was already 65 million in 2000 and attended more than 75 million in 2010. A population that is gathered for more than two-thirds on the shores of the northern Atlantic and Mediterranean region, the life of people in these countries is closely linked to climate and to these changes.

Climate data recorded in the Maghreb region during the twentieth century indicate a warming in this century estimated at more than 1 °C, despite its small share in global emission of greenhouse gases (GHGs) (www.anme.nat.tn). These data also show an increase in droughts and floods. Thus, we started a drought every 10 years at the beginning of the century to 5–6 years of drought in a decade now (Agoumi 2003).

The North African countries suffer from recurrent droughts, whose frequency has increased over the last 40 years. Increasing climate variability affects in particular the center and the north of these countries and the predictions of the Intergovernmental Panel on Climate Change (IPCC 2013) are worrying, even alarming. They indeed show a worsening trend for future decades (OSS 2009). The central Maghreb is characterized by a low average annual rainfall and strong fluctuations in rainfall. In recent decades, he suffered several episodes of drought and the latter tends to be more structural than cyclical. In years of drought, the scarcity of rainfall affects pluvial agriculture which occupies an important agricultural area in the region but also all other sectors dependent on water resources. Irrigated agriculture is also affected and measures of restrictions on water allocation are generally applied (OSS 2009).

Climate data recorded in the region during the twentieth century indicate a warming during this century estimated at more than 1 °C with an accentuated trend in the last 30 years (Fig. 2). These data also show a net increase in the frequency of droughts and floods. Thus, we started a drought every 10 years at the beginning of the century to 5–6 years of drought in a decade now. The general circulation models, even if they are not accurate enough for this region does not have a centered model converge to estimate a likely warming in the region of about 2° to 4° during the twenty-first century.

Trend analysis of rainfall variability in the Mediterranean region shows a significant decrease in rainfall from 1970 (Knippertz et al. 2003; New et al. 2001; Rodrigo and Trigo 2007; Khomsi et al. 2013). This downward trend is higher in winter (Jacobeit 2000; Giorgi 2002). In the Maghreb, Tramblay et al. (2013) observed a negative trend in precipitation intensity, in annual maximum precipitation, and in the 95th percentile of precipitation.

The vulnerability of the climate change in the Mediterranean's south regions varies depending on the part of their economy sensitive to climate (agriculture, tourism, infrastructures, energy, and ecosystem). In Tunisia, agriculture represents 16% of the workforce and 12% of GDP in 2006 (BCT 2007). After a decline of 7% in 2005 caused by drought, its rate of growth was 2.5% in 2006 and 2.1% in 2007.

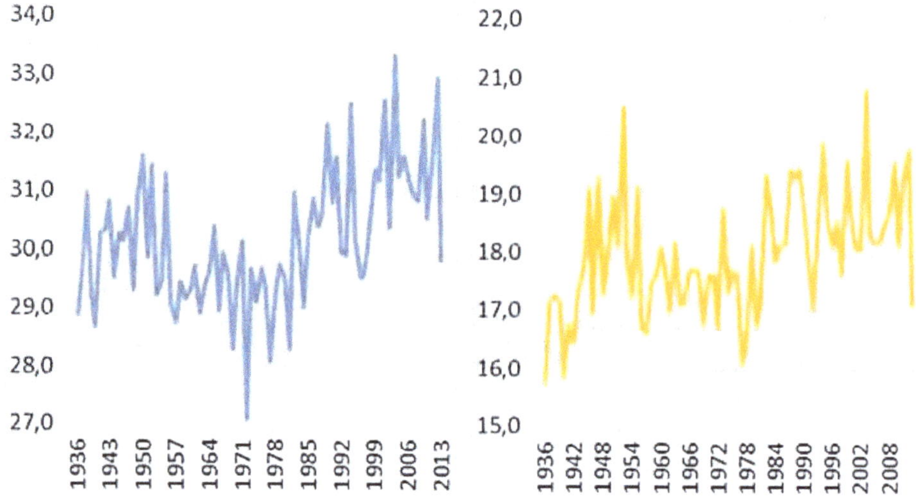

Fig. 2 Evolution of average maximum and minimum temperature in Algiers at Dar El Beida station 1936–2012 (Matari 2016)

For Algeria, agriculture represents 20% of the workforce and 8% of GDP in 2009. For Morocco, agriculture accounts for 40% of the workforce and 17.7% of GDP in 2006. This sector in these countries is directly related to the availability of water which in turn is directly related to rainfall.

Climate Change, Future Projection, and Adaptation

Algeria

Water Potential and Climate Change

Water potential of Algeria is estimated at 19.4 billion m^3/year according to the Ministry of Water Resources. The groundwater resources contained in the northern aquifers of the country (renewable resources) are estimated at 2 billion m^3/year and the surface water is estimated at 12 billion m^3/year. The water reserves of southern Algeria are very important and are of the order of 40,000 billion, this resource is not renewable according to the Ministry of Water Resources.

Water and its management represent a major problem in Algeria following specially in the South. It is amplified by the reduction of rainfall affecting the country since the 1970s. For example, the rainfall, in the northwest, decreases more than 20% (Figs. 2 and 3) and the temperature has increased as shown in Fig. 3. Droughts in rehearsing raise represent the most serious concerns for citizens and managers of the water sector. The climate scenarios and the availability of water are alarming and might have direct and indirect bad consequences on the potential of water resources (quantity and quality) and on their use in the Drinking Water Supply, in agriculture, and in industry.

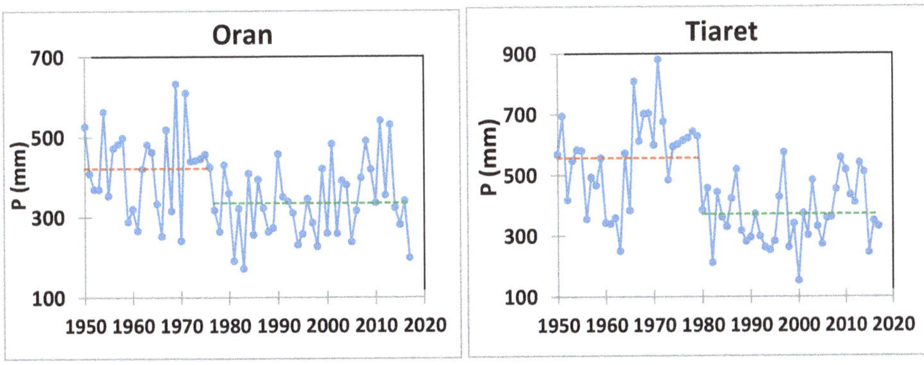

Fig. 3 Trend of annual rainfall variability at Oran and Tiaret stations

Water resources in Algeria are limited, vulnerable, and unevenly distributed in terms of geography. Many dams have been constructed to develop surface water storage capacity, whereas in 1962 there were only 13 dams capable of storing 450 million cubic meters of water, mainly used for irrigation of the plains in the west of the country (http://anbt-dz.com/). The storage capacity has multiplied several times and has reached eighty dams with a capacity of more than eight billion cubic meters in 2017.

According to the Ministry of Water Resources (http://www.mree.gov.dz), drinking water supply, after being confined in big cities, the network connection rate has increased from 78% in 1999 to 98% in 2017. The network is over 127,000 km in 2017. The increase in water availability per capita has increased from 123 l in 1999 to 180 l in 2017, but water distribution systems still lose about one-third of the water distributed due to leaks.

Immediate Water Needs

Water consumption in the agricultural sector is about 7 billion cubic meters per year, which is about 70% of the total volume of water consumed nationally. The drinking water consumption of the population is estimated at 2.5 billion cubic meters per year.

Future Water Needs

Climate change, strong urban and demographic growth, and the nature of water resource management will exacerbate water stress. The demand for drinking water can easily exceed 4 billion cubic meter per year over the next few decades.

In agriculture, we have two possible scenarios:

- **The First Scenario**:

  Represented by a slight change in the current management of water resources, an average increase in irrigated land and the maintenance of the current quality of agriculture, the demand for irrigation water will reach 15.4 billion m^3
- **The Second Scenario**:

  To improve the efficiency of water resources management, increase the irrigated area to 2,000,000 hectares and develop cereal and fodder crops, the demand for irrigation water will increase to 20 billion m^3.

In 1962, theoretical water availability per inhabitant per year was in the order of 1500 m^3, in 1990 was 720 m^3, 680 m^3 in 1995, 630 m^3 in 1998, 500 m^3 in 2009, and it will be only 430 m^3 in 2020.

Algeria is one of the countries where water availability per capita is below the threshold set by the World Bank (1000 m^3/hab/an). Water resources are estimated to 19 billion m^3 (corresponding to about 600 m^3/hab/an). The mobilization of water is facing the following problems:

- The spatial distribution of surface water resources requires substantial transfers to meet the needs of the less fitted regions; these facilities and operations of water mobilization are more and more expensive.
- The effectiveness of reservoir dams is reduced by the high evaporation affecting water bodies (reservoirs evaporate in Algeria from 1.3 to 2.2 m/year and the problem of siltation (32 million m^3/year for 52 dams whose initial capacity is about 5.2 billion m^3).
- The rate of loss in the networks of drinking water varies from 30% to 40%.

In Algeria, the observed climate change across the globe has resulted in:

- An increase in the average, minimum, and maximum temperature recorded on all stations in northern Algeria since the 1970s and that continues to these days.
- On other records, the impacts of climate change have caused our country endemic drought since 1970, resulting in a worsening desertification with a degradation of more than 8% of the forest, wind, and water erosion of soil.
- A trend in rainfall during the past 25 years had a negative impact on water resources by:
- A rainfall deficit of about 30%
- An impact on the level of filling of dams and groundwater recharge
- The impact on the socioeconomic structure
- An impact on the environment (desertification, urban pollution, industrial and agricultural, decline of groundwater levels and salinization of water result of the overexploitation of groundwater and drought)

Impact of Climate Change on Superficial and Underground Water Resources

Indeed, the 90 billion m^3, which precipitate in average each year, only 12 billion m^3 is surface water resources and two-thirds of this volume could be mobilized.

The decrease in rainfall creates a water deficit. It constitutes the most important factors to lead to reduce the flow of the seasons of winter and spring.

For example, lower liquid intake, on an annual scale, at the outlets of five basins in western Algeria (Fig. 7), varies from 16% (Basin of Sidi Ali Ben Youb) to 55% (chouly basins). At Bouhanifia dam, flows have registered a reduction of about 56% since 1979 date of change in the hydrological regime and 55% of reduction at Beni Bahdel dam since 1974 (Fig. 4).

At the level of the plain of Ghriss (west of Algeria) the exploited water varies from 72 to 88 hm^3 /year, while the natural recharge is about 64 hm^3/year (Fig. 5).

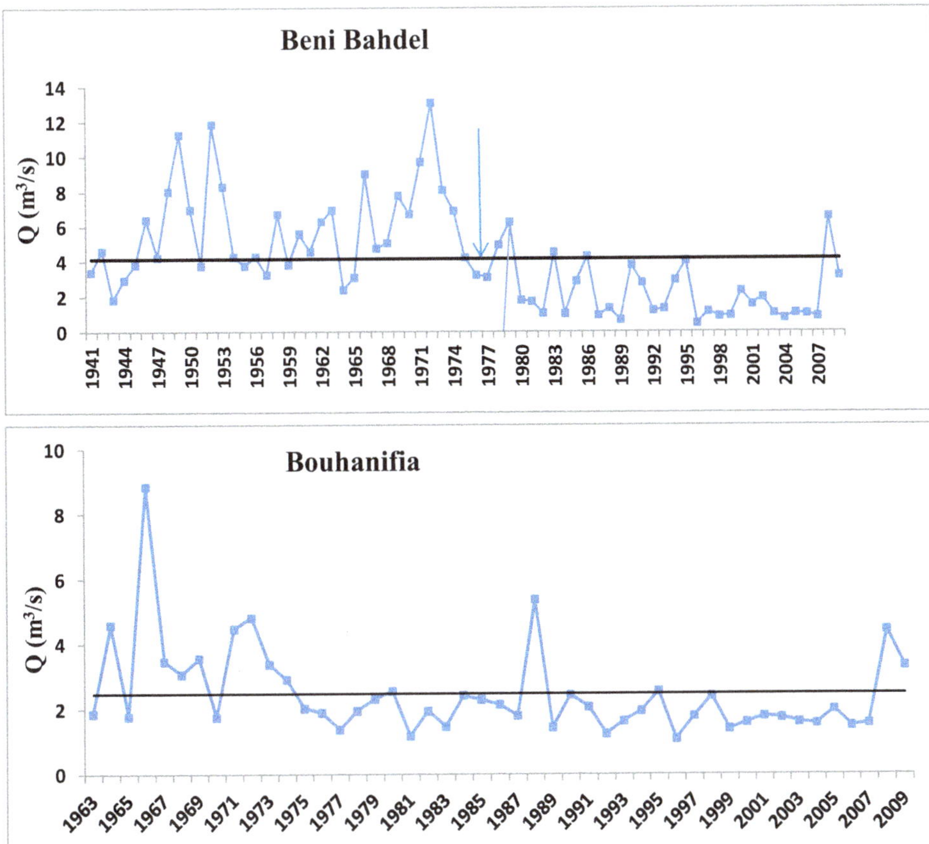

Fig. 4 Inflow to Beni Bahdel and Bouhanifia dams

This decrease is mainly due to the lower rainfall in recent decades. This imbalance has led to a lowering of the groundwater from 5 m to 40 m.

For the Chellif Basin, the total number of wells and drillings increased steadily from 5 in 1850 to 2080 in 2005. This increase in the number of wells and drillings resulted in a change in the amount of mobilized water. These volumes were 3.75 Mm^3 in 1970 and 110.7 Mm^3 in 2005 in the high and medium chellif (Meddi and Boucefiane 2009).

Future Projection

The study of the temporal and spatial evolution of surface and underground water resources is considered as a major decision-making tool for the socioeconomic development of the Maghreb countries.

Dams are filled mainly by the few floods recorded during the rainy season. The recharge of the aquifers is ensured by infiltrations which occur in winter and spring. This recharge has become insufficient since the mid-1970s following the decrease in rainfall (Meddi and Hubert 2003; Hallouz et al. 2013; Zeroual et al. 2017) and the

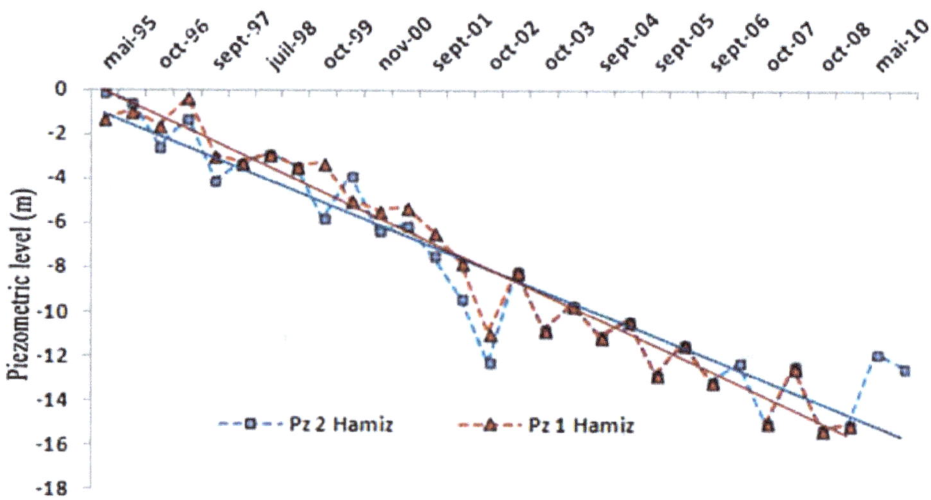

Fig. 5 Evolution of the piezometric level in Pz1 and Pz2 Hamiz, plain of Mitidja (1995 to 2011)

overexploitation of groundwater to meet irrigation water and drinking water needs. The drop in piezometric levels reached 40 m for the Mitidja Plain (Demmak 2008).

Climate projections will make it possible to establish scenarios for the availability of water resources at future horizons. Projections of the spatio-temporal variability of precipitation and temperature for the periods 2006–2060 and 2045–2100 performed by Zeroual (2017) using the RCA4 model (NorESM1-M), scenarios RCP4.5 and RCP8.5, for the periods 2006–2060 and 2045–2100 showed a stabilization of average rainfall compared to the period from 1951 to 2005 in eastern and southern Algeria and a decrease in the western part of Algeria (Figs. 6 and 7). This decrease will change from RCP4.5 to RCP8.5 and from 2006–2060 to 2045–2100. The climate models predict a decrease in precipitation and an increase in temperatures both on annual and monthly scales for the periods 2006–2060 and 2045–2100 (Zeroual 2017).

The impact on the flows will be very important as shown by Kahlerras et al. (2018) in a work on the Mazafran Basin (Fig. 8), in the center of Algeria. The Oued Mazafran, which drains a basin of 280 km^2, for the RC4.5 scenario, will record a decrease in flows will be of the order of 29% in 2033 to reach a volume of 134.01 Mm^3 compared to the reference period (1998–2014) which recorded an average volume of 189.85 Mm^3. For scenario RC8.5, a decrease will be of the order of 31.6% in 2028 with a volume of 129.8 Mm^3, by 2040, the volume will be of the order of 129.9 Mm^3. These 2 years will be the least rainy with 178.4 mm/year and 184.8 mm/year, respectively (Kahlerras et al. 2018).

Kahlerras et al. (2018) have shown that the demand for water for the Mazafran Basin, whose population is estimated at about 615,956 inhabitants in 1998 and will be of the order of 1,724,880 inhabitants by 2050, there will be a recovery of demand until 2040, the deficit after this date will be of the order of 3.44 Mm^3 in 2045 and will

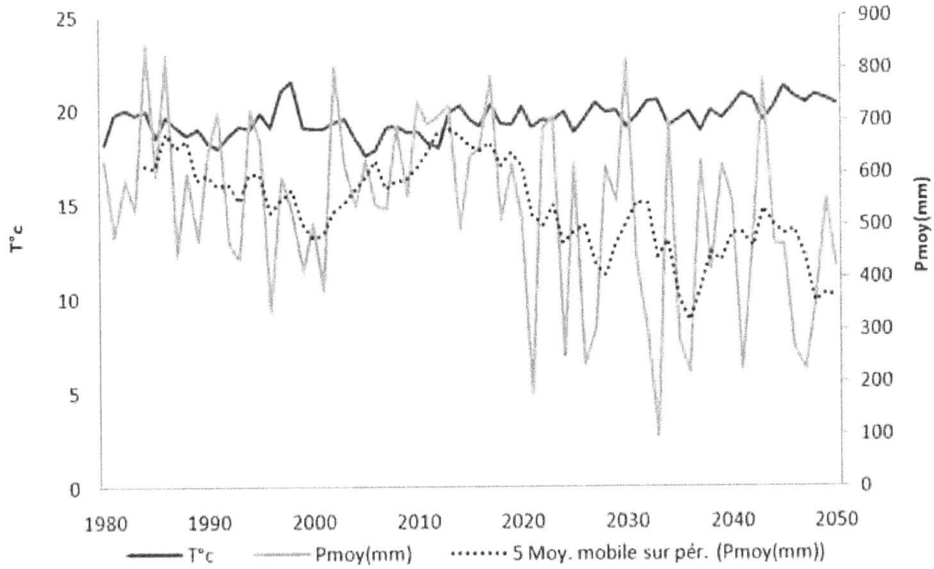

Fig. 6 Projected temperature and rainfall trend change at the time horizon of 2050 (scenario RC4.5) at the Mitidja plain (Kahlerras et al. 2018)

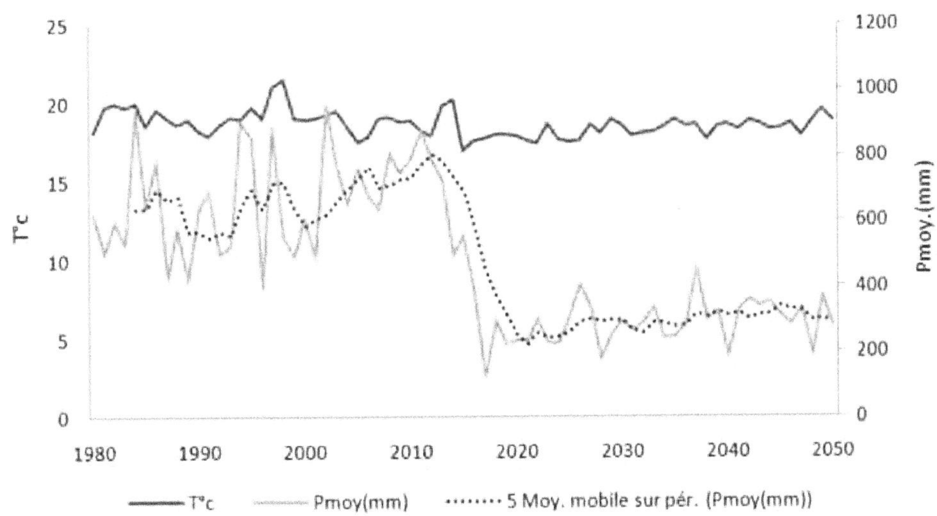

Fig. 7 Projected temperature and rainfall trend change at the time horizon of 2050 (scenario RC8.5) at the Mitidja plain (Kahlerras et al. 2018)

reach 95.3 Mm^3 in 2050 for scenario RC4.5 (Fig. 9). The situation will be more critical for the pessimistic scenario (RC8.5) where the rupture between supply and demand will be recorded as early as 2025. The deficit will be of the order of 56.65Mm^3 in 2030 and will reach 130.95 Mm^3 in 2050 (Fig. 10).

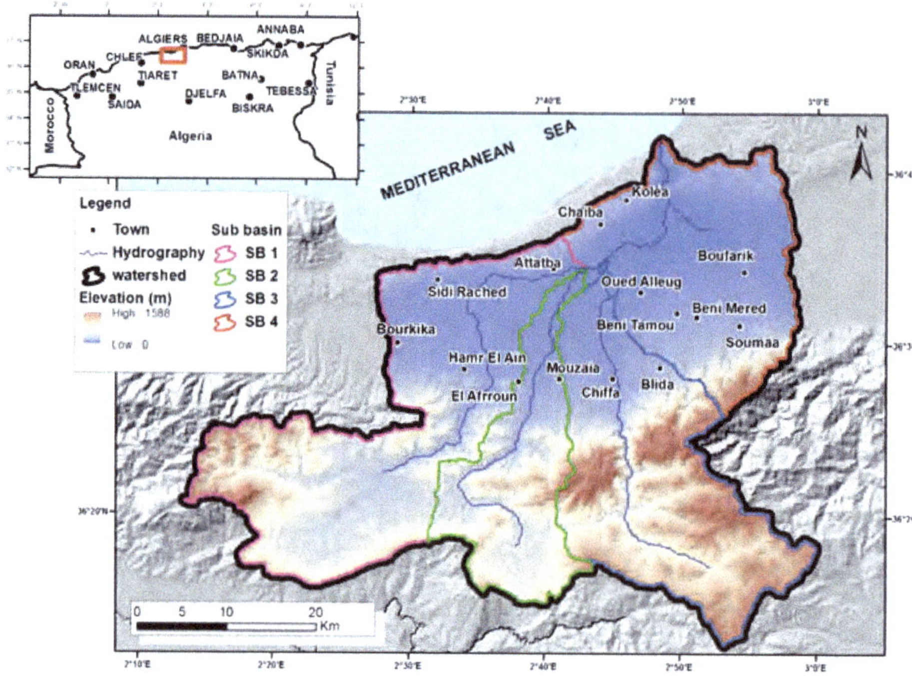

Fig. 8 Geographical situation of the Mazafran watershed (Kahlerras et al. 2018)

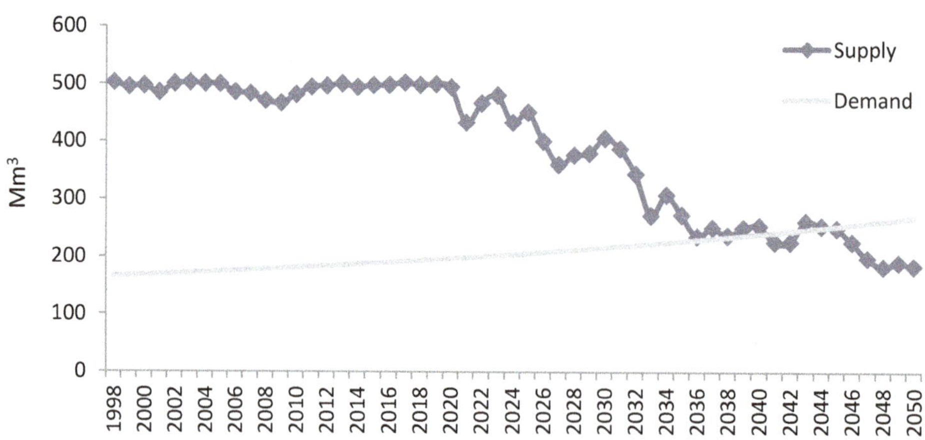

Fig. 9 Assessment of water resources management for the RC 4.5 scenario (Kahlerras et al. 2018)

For basins located in western Algeria, the region most affected by the reduction in rainfall by 25% (Meddi and Hubert 2003), winter flows will show a clear downward trend, mainly due to the decrease in rainfall and the increase in PET for the two scenarios RCP8.5 and RCP4.5. On the other hand, spring flows will increase according to RCP 8.5 and decrease according to RCP 4.5. In terms of observed differences, the RCP 8.5 scenario forecasts a decrease in winter flows of about 33% on average in all the studied basins by 2039, 31% by 2069, and 50% by the end of

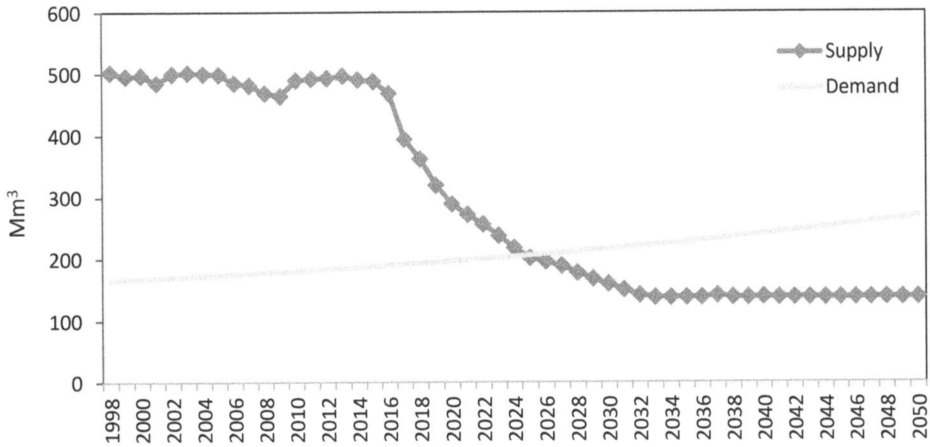

Fig. 10 Assessment of water resources management for the RC 8.5 scenario (Kahlerras et al. 2018)

the twenty-first century. The RCP4.5 scenario predicts a decrease of about 50 to 60% for all horizons. In addition, the decrease is estimated at 80% for spring flows according to the RCP4.5 scenario. In contrast, the RCP8.5 scenario forecasts a 25% increase for the same season (Haddour et al. 2019).

Palliative Measurements

To cope with the change in rainfall reduction, Algeria has undertaken a major program based on large transfers and the use of the non-convent water.

The state has made a considerable effort to increase the connection rate, mainly in rural areas. This rate increased from 44% for some rural regions to more than 60% (http://www.mre.gov.dz). The recent decision to use non-conventional water, desalination, will provide a substantial addition to alleviating the suffering of populations of large cities and rural areas. An emergency program was decided by the government for commissioning 21 desalination stations with a total capacity of 57,500 m^3/day (http://www.mre.gov.dz). For this project, the Government aims to strengthen the water supply of some urban areas situated in the coasts in the summer season. Water dams originally planned for these towns will be transferred to the interior cities. This procedure will develop these towns and villages that are in poverty and enable people to develop agriculture and prevent diseases inherent to the lack of water.

During the warm season, the deficit is growing due to the increasing demand and declining reserves of conventional water (http://www.mre.gov.dz). However, these programs cannot alone solve the water problem in the long term. Rigorous management and increased storage capacities of surface and groundwater are essential to meet the demand. The reuse of wastewater, especially in aquifer recharge and irrigation, will increase, with no doubt, the water resource available in the medium and long term.

There is a considerable waste in agriculture. To remedy to this, it is necessary to restructure the peasants in unions capable to organize the operation of irrigation and to modernize, as appropriate, irrigation techniques.

For a better management, decentralization of the management of water supply and sanitation in order to promote the emergence of effective management by elected authorities and representative committees of citizens. We must encourage and facilitate management by major river basins (basin agencies).

Once the impacts on water resources will be determined for the main socioeconomic activities, strategies to adapt to the negative impacts of these climate change scenarios for Algeria can be developed for the benefit of policy makers. They will be able to draw a proper water policy in the medium and long term.

Large Transfers

To cope with water shortages in certain parts of the country, the government has started the constructions of several large transfers across the country, we give as examples:

- Beni Haroun Development in the east of the country: 242 million m^3 will be used to supply drinking water for 4.62 million inhabitants of several department: Jijel, Mila, Oum El Bouaghi, Batna, Constantine, Ain M'lila, and Khenchela. Also, 262 million m^3 for irrigation of 30,000 hectares.
- Mobilization of water resources in Algiers: This project consists of projects set to ensure an annual volume of 595 million m^3. A total of 435 million m^3/year for drinking water supply for 7.95 million inhabitants and 160 million m^3 for the irrigation of about 30,000 ha.
- Mobilization of water resources in western Algeria: The development of the water production of Chellif-Kerrada called MAO provides 155 million m^3/year for the drinking water in the corridor Mostaganem-Arzew-Oran.

Desalination of Seawater

A very ambitious program was initiated to meet the water demand constantly increasing through desalination of sea water. This resource is inexhaustible and can be exhausted on a coast of 1200 km. In addition, the population and the industries that are large consumers of water are near the sea to defend this approach. The field of desalination of seawater has in recent years a remarkable technological step through the development of different processes. The program is composed of large monoblock stations and a desalination plant. For the first category, 21 stations with a total capacity of 57,500 m^3/day:

- 13,000 m^3/day for the western region.
- 22,500 m^3/day for the central region
- 7000 m^3/day for the east region

For the large stations, the regional distribution is made as follows: western region (1.29 million m^3/day), the central region (0.45 million m^3/day), and the eastern region (0.15 million m^3/day).

Recommendations

- Algeria will build more than 40 dams to reach 130 dams by 2030 to increase the mobilized water resources.
- Reduce leakage from urban and rural drinking water systems by 30% to 20% by 2030.
- Save between 20% and 25% of water intended for irrigation by 2030.
- Economy of 20% (1.4 billion m^3) in agriculture could provide drinking water to half of the Algerian population
- The used water treatment stations can process up to one billion m^3 in the future. This unconventional water resource is an excellent addition to irrigation and can easily cover more than 100,000 hectares.
- Integrated Water Resources Management (IWRM) is a good solution for smart and optimal management through the development of scenarios to counter the impacts of future climate change.
- Develop integrated management of water resources by basin.
- The development of research projects to be conducted by a consortium of farmers and research laboratories to develop optimized complete solutions for the reuse of treated water in irrigation.

Morocco

Water Potential and Climate Change

Morocco has a potential of water estimated at 22 billion m^3, 18 billion m^3 of surface water, and 4 billion m^3 of groundwater (Choukr-Allah 2011). Morocco counted 135 large dams with a total capacity of 17,500 million m^3 in 2009, as well as about 100 small dams with a capacity of 100 million m^3 for local needs in drinking water, irrigation, and livestock, and there are also 13 systems for transferring water between watersheds (https://water.fanack.com/morocco/water-infrastructure-in-morocco/). The national water strategy foresees the construction of about 60 large dams by 2030, with a total capacity of about 7 million m^3 plus 1000 small dams for local development (Alaoui 2013). The water resources reduce in particular because of overexploitation (Fig. 11). This depletion will increase with the advent of climate change. Similarly, the development of industrialization and intensification of agriculture water consumption would cause further degradation of groundwater.

In addition, water resources will be subject to further exploitation, the domestic and industrial wastewater will constitute a problem concern threatening the human health and the environment. Thus, the urban wastewater was about 600 Mm^3/year in 2015, it will increase to about 750 Mm^3/year in 2030 and 900 Mm^3/year by 2050 (World Bank Group 2017b).

The recorded rainfall shows a negative national trend (Fig. 12). The spring rains have fallen over 40% since the 1960s. The drought seems to become more persistent over time. The temperature has increased as shown by the Fig. 13. The dry maximum length is rising during the rainy season, significantly during the end of this season (February to April) where it increased 15 days since the 1960s. At the same time, the

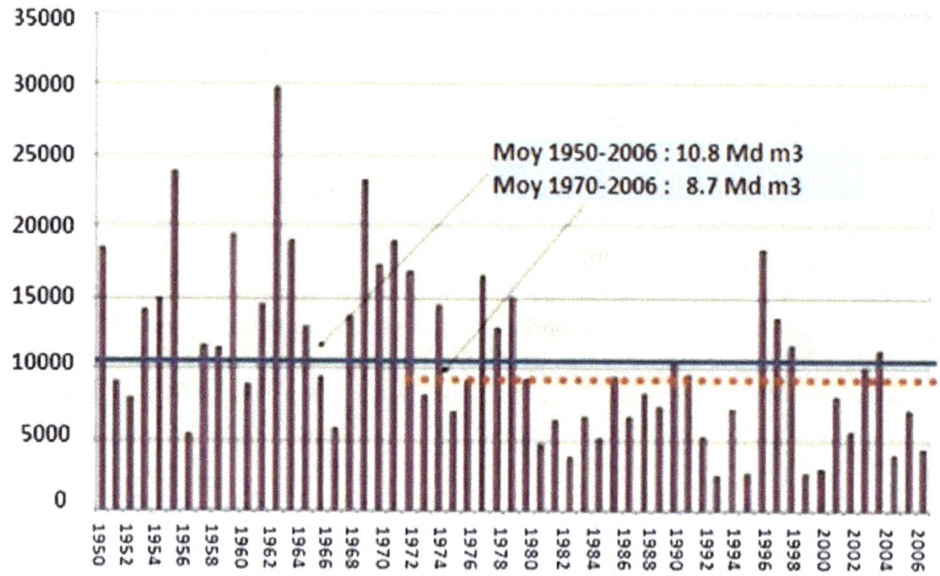

Fig. 11 Evolution of the water contributions in dams of Morocco (Ben Abdelfadel and Driouech 2008)

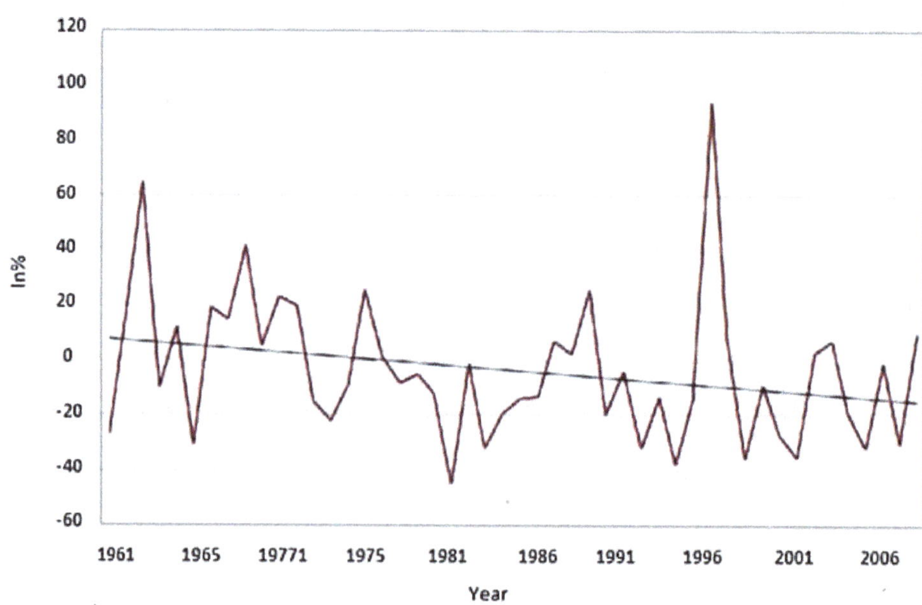

Fig. 12 National averages (%), defects of the cumulative rainfall totals annual, calculated on the 14 meteorological stations in Morocco (1961–2008), Driouech (2010)

total number of rainy days shows a negative trend and indicates an increase in the annual number of dry days. Many regions have become drier between 1961–1985 and 1986–2005 (Ben Abdelfadel and Driouech 2008).

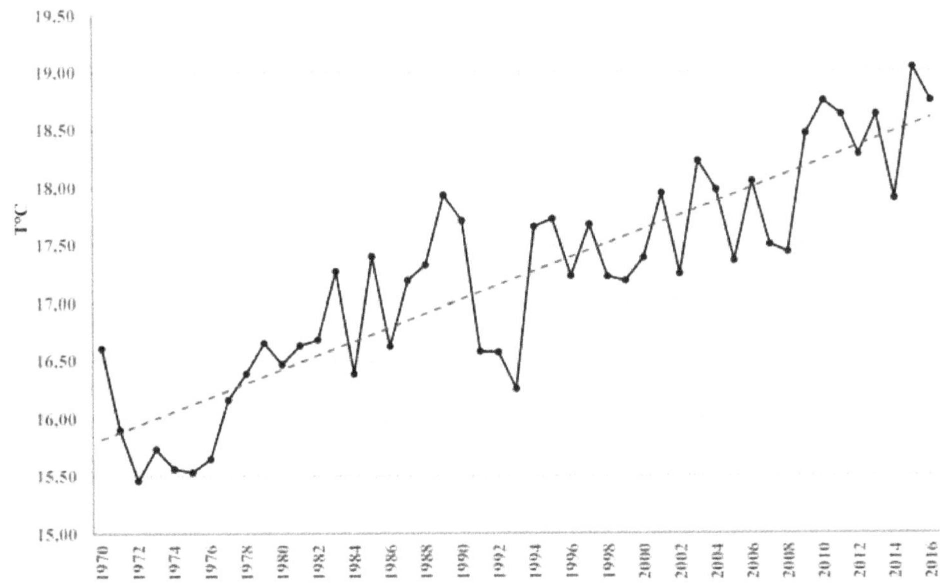

Fig. 13 Average annual temperatures at the Oujda station in Morocco (El Hafid et al. 2017)

The drought has led the agricultural sector in Morocco, for example, to limit the not needed annual crops, to the prohibition of any new tree planting and to the prohibition of vegetable crops in dry years from 1981 to 1982.

The potential of renewable water resources is estimated at 22 billion m^3 per year of which 18 billion m^3 of surface water. From the available potential two-thirds are currently mobilized and used up to 88% in irrigation and 12% in the drinking water.

Morocco's Third National Communication on Climate Change 2016 has shown that, according to observed climate trends and long-term projections, climate change is already underway based on the climate trends observed between 1960 and 2005 (MDCE 2016). Average annual temperatures have increased from 1.0 to over 1.8 °C (MDCE 2016). Precipitation has declined between 3% and 30% with a 26% decline in the northwestern region of Morocco (MDCE 2016). On average, warming varied between 1 and 3 °C depending on the region during the period 1998–2007 compared to 1971–1980 (MDCE 2016).

Surface Water

According to the second submission, river flows have recorded a 20% reduction since 1950 (SCNCCRM 2009). Knowing that the surface water resources on the whole territory are evaluated in average year to 18 billion m^3, varying, according to the years, from 5 to 50 billion m^3. More than 51% of the surface water resources are produced at the level of four hydrographic basins (Loukkos, Tangier, Mediterranean Coastal, and Sebou) (SCNCCRM 2009).

The last four decades have seen a marked decrease in stream flows and natural recharge of groundwater (Sinan and Belhouji 2016). The decreases in superficial

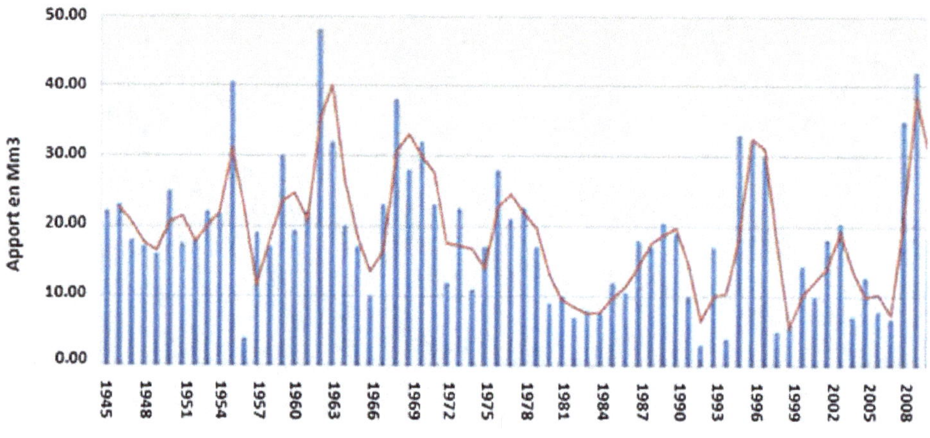

Fig. 14 History of surface water supplies in Morocco 1945–2010 (MDCE 2014 in Sinan and Belhouji 2016)

flows reached nearly 20% between 1970–2006 and 1950–2006 and exceed 70% in certain regions of Morocco (Sinan and Belhouji 2016), Fig. 14.

To show the probable effects of climate change on surface water resources, two basins feeding the dams of Hassan Addakhil (located on Wadi Ziz in the south-east) and Idris I (located on Wadi Innaouene, tributary of the oued Sebou, central Morocco) are considered. In Morocco, like other countries around the Mediterranean, surface water flows depend mainly on rainfall.

Figs. 15 and 16 present the history of water inflow in the Hassan Addakhil (by Wadi Ziz) and Idriss I (by Wadi Inaouene, Basin of Sebou) dams over the period 1950–2002 (Sinan et al. 2009; Sinan and Belhouji 2016; MDCE 2016). They show a clear downward trend in these inputs, especially since the beginning of the 1980s (more than 30 years ago), a clear consequence of the negative effects of climate change. The average water inflow during this period (1950–2002) was respectively 122 Mm^3/year and 516 Mm^3/year for the Hassan Addakhil and Idriss I dams (Sinan et al. 2009; Sinan and Belhouji 2016).

The water resources of the two basins will record decreases as below compared to the end of the twentieth century (Sinan and Belhouji 2016):

- 2020: −7.6% (optimistic scenario) and −8.6% (pessimistic scenario)
- 2050: −2.3% (optimistic scenario) and −12.8% (pessimistic scenario)
- 2080: −7.6% (optimistic scenario) and −40.6% (pessimistic scenario)

Groundwater

Groundwater represents about 20% of the potential of Morocco's water resources. These waters are of great importance for the socioeconomic development of the Kingdom. Morocco has 96 aquifers, 21 of which are deep and 75 shallow and cover a total area of 80,000 km^2. The level of these aquifers shows a continuous decrease due to drought, reduction of natural recharge, and overexploitation. This reduction in groundwater level is about 2 m each year. The volume of groundwater withdrawn

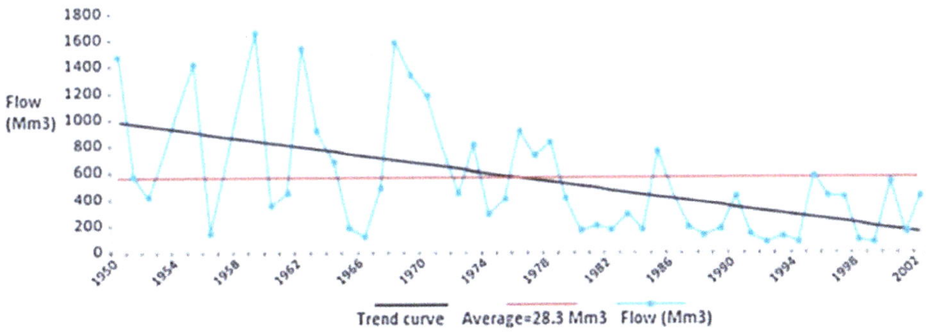

Fig. 15 Historical water inflow in Hassan Addakhil Dam by Wadi Ziz (Sinan et al. 2009)

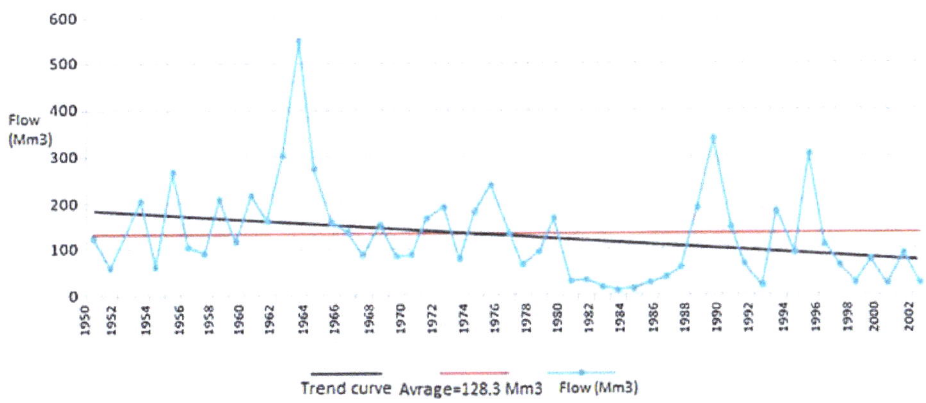

Fig. 16 History of water supplies in the Idriss I dam by the Inaouene wadi – Sebou basin (Sinan et al. 2009)

from the aquifers exceeds the renewable water resources. In fact, recent estimates estimate withdrawals at 5 billion m^3/year, which is a reflection of the nonrenewable potential of groundwater with an average volume of 1 billion m^3/year. This over-exploitation has led to a general decline in groundwater levels sometimes exceeding 2 m/year in some aquifers (Sinan and Belhouji 2016).

The impact of reduced rainfall and consequently the groundwater recharge, as well as the overexploitation of groundwater has led to the lowering of the piezometric level. According to Amraoui (2019), between 1980 and 2017, the deep aquifer of Saiss plain at the level of the piezometer 290/22, controlling the region of Meknes, shows an average decrease of 3 m/year (Fig. 17). Groundwater exploitation in this area increased from 4 million m^3 in 2005 to almost 10 Mm in 2014 (Ameur 2017). For this aquifer, the Sebou Watershed Agency (ABHS) has identified more than 9000 water points exploiting the water table, most of which are not authorized (Amraoui 2019). This is the equivalent of 4 holes per Km^2 (Amraoui 2019). In terms of use, agriculture would take 225 million m^3 annually and 100 million m^3 of drinking water (Amraoui 2019).

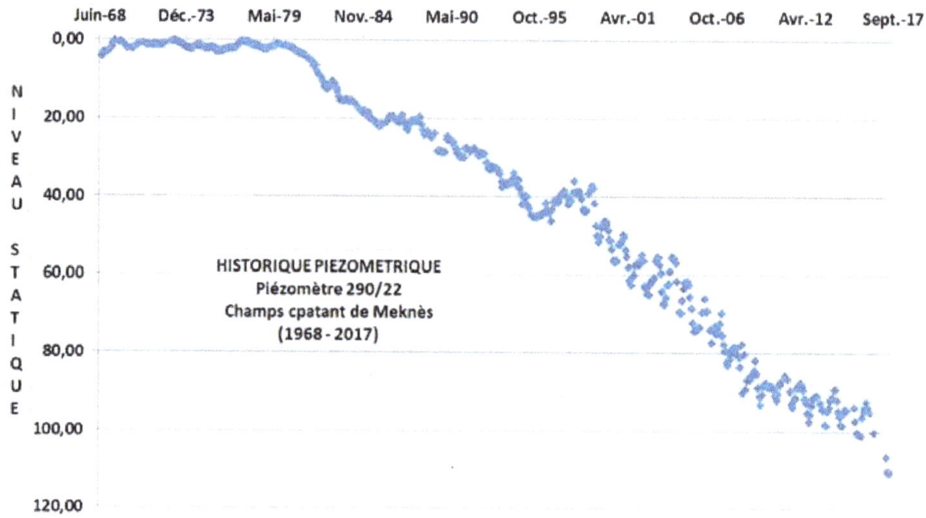

Fig. 17 Decrease of the level of aquifer of Siass plain, Morocco (Amraoui 2019)

The impact of climate change and reduction of water resources have led to:

- Worsening of water supply in rural areas
- Very important economic impact on the agricultural sector: drop in grain production, irrigation water deficit, increased unemployment in rural areas
- The impact on water quality

Future Projections

Projections of climate to future horizons indicate a negative trend of annual rainfall that varies between 10 and 20% to reach 30% in Saharan regions by 2100. The trend in annual average temperatures show an increase of 0.5 to 1 °C in 2020 and 1 to 1.5 °C in the 2050 and 2080 horizons across the country (MDCE 2016; Sinan and Belhouji 2016). The observed temperature has increased by 0.5 °C each decade since 1970. It is above the world average of about 0.15 °C (World Bank Group, 2017). The forecasts show that Morocco will reach 3 to 7 °C from here at 2100, with 4 to 7 °C for the summer months (June, July, and August). Annual rainfall is expected to decrease by 10 to 40%, especially by 10 to 30% during the rainy season from October to April and by 10 to 40% during the dry season from May to September (World Bank Group 2017a).

The water potential in Morocco is estimated at 22 billion cubic meters per year. With a forecast to 38 million inhabitants in 2030 and despite a considerable effort, in terms of construction of hydraulic structures and access to unconventional resources, this quantity of water could fall to 500 m^3 by 2030 under the pressure of climate change, population growth, and pollution increase in economic activity (INDC 2015). Over the next 20 years, Morocco will be more vulnerable to drought and other extreme weather events (World Bank Group 2017b).

These projections also show that the water/year/inhabitant capital would be significantly reduced by 2050 and 2080, thus creating a situation of water scarcity (water capital/year/inhabitant of less than 500 m^3/year/inhabitant) between the 2030 and 2050 horizons for all considered socioeconomic scenarios (Sinan and Belhouji 2016; MDCE 2016), Table 1 and Fig. 18.

Palliative Measurements

The adaptation is a process allowing improving, developing, and implementing strategies to mitigate the consequences of climatic events, to cope and to take advantage.

- Participatory approach: associate all communities and socio-professional categories capable to suffer the consequences of climate change
- Proactive approach: situations of crisis, so far as it is possible
- Optics of sustainable development
- Adaptation of infrastructure
- Use of nonconventional water resources

Currently, Morocco has 144 large dams and 13 others under construction. As part of a National Water Plan by 2030 and to meet the water needs of the population to avoid an imbalance between demand and supply by 2030, Morocco will realize close

Table 1 Projection of capita m^3/inhabitant/year (Sinan and Belhouji 2016)

Year	Optimistic scenario	Pessimistic scenario
1960	2560	2560
2000	775	775
2020	575	569
2050	518	464
2080	342	217

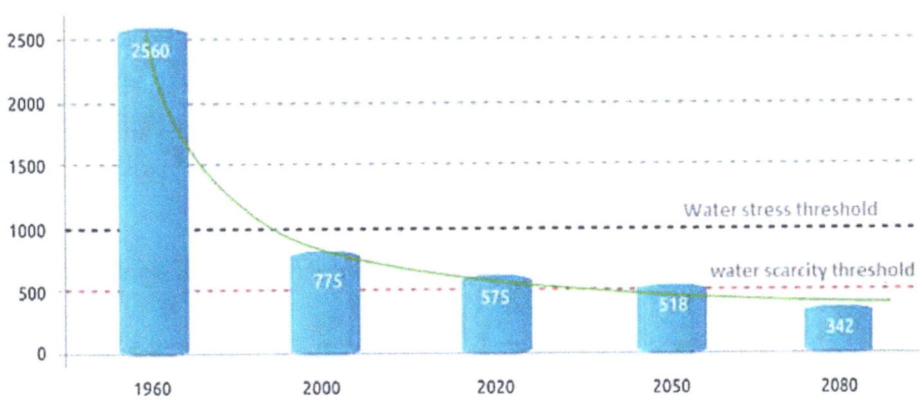

Fig. 18 Water capital projection (m^3/ inhabitant /year) optimistic scenario (Sinan and Belhouji 2016)

to 1000 small dams and dams in different regions of Morocco between 2019 and 2026 (Plan National de l'Eau, horizon 2030).

Nonconventional Water Resources

(a) **Reuse of Treated Wastewater**:

According to the national water plan for 2030, the potential of waters that can be reused by 2030 is around 325 million m^3/year. This volume constitutes the potential of the usable water. To achieve these objectives, Morocco plans to carry out 162 projects in the watersheds of: Loukkos, Moulouya, Sebou, Bou Regreg Chaouia, Oum Er Rabiî, Tensift, Souss Massa Drâa, Guir Ziz Rhéris, and Sakia El Amra Oued Eddahab.

(b) **Desalination**

Desalination of seawater is an alternative For Morocco. Morocco is using since 1990 the desalination of seawater including the establishment of a station in 1995. This unit has a production capacity of 7000 m^3/day of drinking water. In order to develop this sector, the government implements in Agadir another desalination station of 80,000 m^3/day, which should be operational in 2010. In addition, in view of saving water, the study suggests to "fight against failures in water networks and water distribution, to raise awareness on water scarcity and promote the use of plants resistant to water stress." Because, the study conducted by the High Commissioner for Planning shows that Morocco loses 1.3 billion per year including 1.1 billion m^3 in agriculture and 200 million m^3 in the drinking water and industrial sector.

A seawater desalination plant will be built in the Agadir region. It will be powered by solar energy, so it is a 100% ecological process in terms of renewable energy. Ultimately, it plans to produce about 275,000 cubic meters of water per day, of which 150,000 will be dedicated to human use and 125,000 to be used for irrigation of the 13,600 hectares of plantations in the southwest of the Kingdom, the maximum production capacity being 450,000 m^3 per day.

Recommendations

According to the national water plan for 2030 of Kingdom of Morocco and to ensure the balance between supply and demand, the following operations will be carried out:

- The increase to 80% of the national average of the efficiency of drinking water distribution networks by 2025.
- The maintenance of the same level until the year 2030 and the search and repair of leaks, which will reduce the demand for drinking water by about 120 million m^3 per year by 2025.
- Saving water in irrigation by transforming traditional irrigation systems to local irrigation system (drip system) at a rate of 50,000 ha per year and saving 4.1

billion m^3 by 2020, and as part of the Green Morocco plan, efforts will be made to save an additional 3.2 billion m^3 by 2030.

- The construction of fourteen dams in progress, a further 35 are planned, in addition to the use of non-traditional water resources, such as the desalination of seawater at a rate of 510 million m^3 and the reuse of 325 million m^3 of wastewater after purification (Plan national water supply by 2030)
- The World Bank (2017) indicates that 35% of potential water resources are wasted, which requires a new strategy to preserve this resource.
- The National Water Plan provides, to meet the demand for water by 2030, an investment of 220 billion DH by 2030, which will be used among others to the construction of 3 dams per year.
- Develop integrated management of water resources by basin.

Tunisia

Water Potential and Climate Change

Tunisia has about 4766 million m^3 of exploitable resources for the year 2017, the equivalent of 450m^3/year/inhabitant for all uses combined. The water mobilized in Tunisia is divided into surface waters (55%) 2630 Mm^3 and groundwater (45%) 2136 Mm^3. Also, it should be noted that 88% of groundwater has salinity greater than 3 g/liter (Lahacher and Pillet 2008). Water resources in Tunisia are distributed as follows: 80% for irrigation, 14% for water drinking, and 6% for industry and tourism (SIGMA 2017). Tunisia had 17 dams before 1990. It built 20 dams during the first ten-year plan between 1990 and 2000. Then, during the 2nd ten-year plan from 2001 to 2011, Tunisia made 10 dams. The current total capacity of the 33 dams is 1.8 Mm^3 in 2018 according to the Ministry of Agriculture. The 253 collinear dams have a total capacity of 266 Mm^3 (Ouasli 2015). The 893 collinear lakes have a total capacity of 93 Mm^3. Groundwater is produced by 138,000 shallow wells and 5400 deep wells (SIGMA 2017). The mobilization of water resources is about 95% of available water resources (Ouasli 2015).

By studying the evolution of the contributions to the main dams over the period from 1947 to 2004, two main conclusions can be distinguished (TSNCUNFCCC 2013):

- The contributions decrease by 27% during the period from 1976 to 2004 compared to the period 1947 to 1975.
- Contributions at the level of the northern dams during the period from 1976 to 2004, on the other hand, are 21% higher than the current average inputs.

Tunisia is located in the category of the least endowed countries with water resources in the Mediterranean. Indeed, for a population of 10 million inhabitants, the average allocation per capita per year is estimated at 450 m^3. This figure will decrease gradually to 350 m^3 per capita in 2030 for a population of 12 million inhabitants (Hamza 2007).

The surface water resources are highly variable in time and space. The average interannual surface water inflow is estimated at 2.7 billion m^3 per year (55.5%), 80% are from northern regions of Tunisia (Elloumi 2016). The groundwater resources are estimated, annual averages, about 2165 billion m^3 (45.5%) (Elloumi 2016). The volume of water mobilized in 2015 reached 4. 6 mm (98%) (Elloumi 2016).

The increase in temperatures would vary between 0.8 and 1.3 °C in 2020 and 2.1 °C in 2050 with the decrease of rainfall by 10 to 30%, the loss of 28% of conventional water resources, and 20% of arable land by 2030.

Nouaceur (2016) divided the rainfall trend for Tunisia into three periods:

- From 1970 to 1986, an absence of trend. This period is characterized by an important variation of the extreme years.
- The second period (1987–2003) is marked by a dry trend that lasts 16 years. The years of the dry and very dry class totaled 55% of the observations for all the stations.
- The third period (2003–2013) is described as wet. The wet and very humid years total almost 48% while the dry and very dry years 31.3%.

Climatic data show that average annual temperatures have increased by about 1.4 °C during the twentieth century, where the warmest temperatures have been recorded since the 1970s. Local warming rates may be higher, for example, in Tunis, temperatures rose by around 3 °C during the twentieth century (World Bank 2013). Annual totals of rainfall have decreased by 5 percent per decade in northern Tunisia since the 1950s (Fig. 19).

The study on the inflow to major dams during the period from 1947 to 2004 revealed two major conclusions: a decrease in discharge of the period 1976–2004 compared to the period 1947–1975 about 27% (Lahacher and Pillet 2008).

Deep and shallow aquifers exploitation increased by 333% and 228%, respectively, between 1968 and 2010. The largest increase occurred before the 1980s (Trabelsi 2016), Fig. 20.

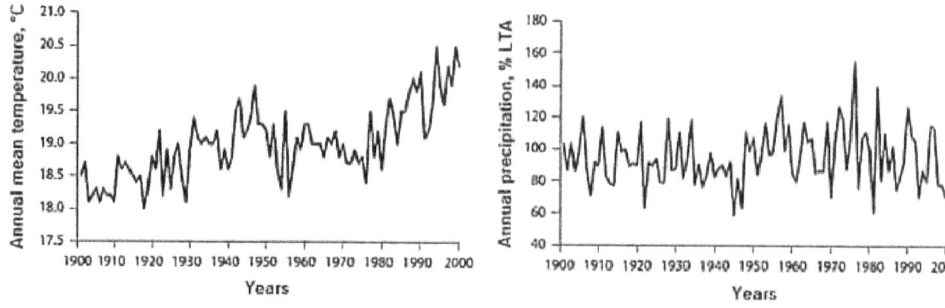

Fig. 19 Average temperatures of the twentieth century and precipitation in Tunisia (Mitchell et al. 2002)

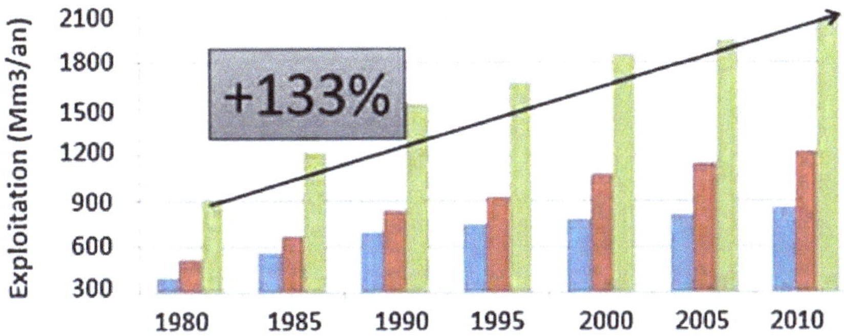

Fig. 20 Evolution of the groundwater exploitation in Tunisia in Mm^3/year (Trabelsi 2016)

The droughts will result from the intensive exploitation of groundwater. The simulation based on the historical evolution of the quality of groundwater resources has led to a decrease of the exploitable groundwater potential equal to 467 m^3 in 2030. This quantity corresponds to the volume of water currently overexploited. For deep waters, the reduction will be 517 million m^3. The exploitation of deep groundwater will require more energy to pump water (Lahacher and Pillet 2008).

The pressure on water resources will be major. Groundwater, coastal, non-renewable aquifers will decline from 28% in 2030, surface water will decrease by 5% on the same horizon, and the summer rainfall will increase the lack of soil moisture (Sarra 2009).

Future Projections

The projections from the HadCM3 model, according to the A2 scenario, show an increase in mean temperature of 1.1 °C by 2020 and 2.1 °C by 2050 compared to the 1961–1990 reference period. The largest increases will be in the South, with 1.3 °C by 2020 and 2.7 °C by 2050 (TSNCUNFCCC 2013). By 2100, projections of average monthly temperatures show a stronger upward trend over almost regions, the expected increase in temperature is above 2 °C (Fig. 21). The maximum is expected in the region of Gafsa in southern Tunisia with an increase of 3.91 °C (Ouasli 2015).

Dry years will become more frequent and intense by 2030, and extremely dry and wet seasons will vary from season to season (TSNCUNFCCC 2013).

The HadCM3 projections for the A2 scenario show a general decrease in average precipitation. This reduction is relatively moderate by 2020 but will become significant by 2050. In 2020, the decrease in annual rainfall varies between −5% in the north and −10% in the extreme south (TSNCUNFCCC 2013).

The decrease in rainfall varies between (−10%) in the northwest and (−30%) in the extreme south of Tunisia by 2050. Three regions can be distinguished: the extreme west of the country is characterized by a slight decrease in the order of (−10%), the south with a significant decrease of about (−29%) compared to the third

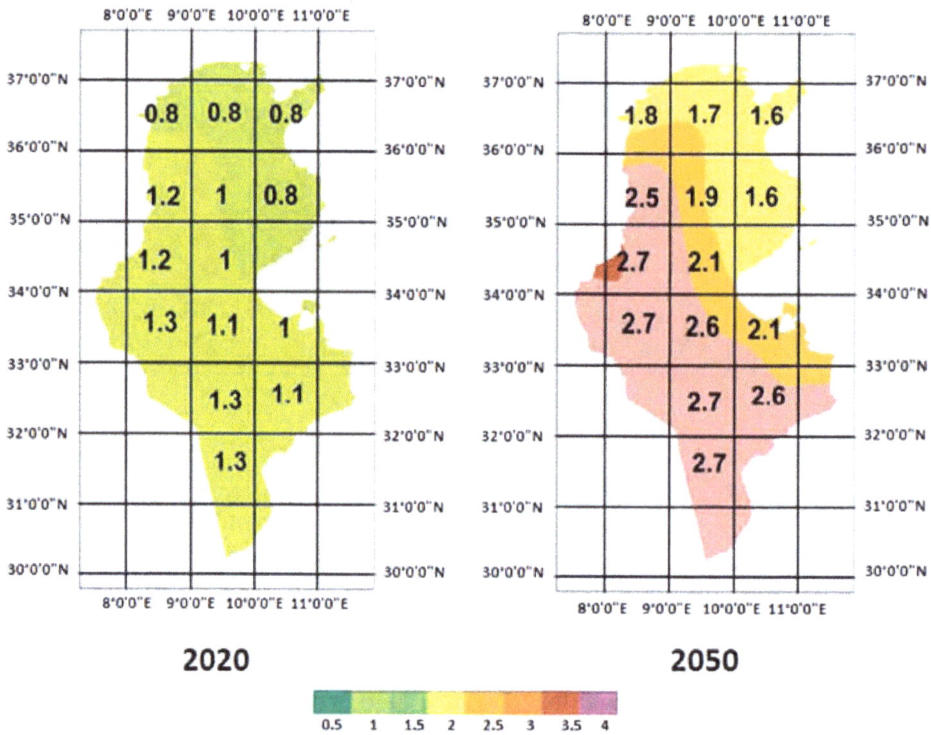

Fig. 21 Temperature (°C) average annual increases in the HadCM3 model (A2 scenario) compared to the reference period in 2020 and 2050 (MARH 2007)

region which includes the rest of the country where the reduction varies between −12% and −16% (TSNCUNFCCC 2013) (Fig. 22).

For seasonal changes by 2020, the lowest reduction in precipitation will be recorded in winter (0% to −7%). The largest drop from −8% to −40% will occupy the space from north to extreme south in summer (TSNCUNFCCC 2013).

The situation in autumn and spring is intermediate with rainfall decreasing from −6% to −12% in the extreme south. Seasonal changes are about the same by 2050 (TSNCUNFCCC 2013).

By 2100, future trends in monthly average rainfall are down for all Tunisia (Fig. 23). The maximum decrease is for the Sfax station, located in North East Tunisia, with a percentage of 78.23%. Projections show a strong gradient between trends in the north and those in the south (Ouasli 2015).

Climate projection models show that conventional water resources will decrease by about 28% by 2030. This decrease will mainly concern shallow groundwater with high salinity, coastal aquifers, and nonrenewable aquifers. In particular, the reduction of surface water reserves will be around 5% by 2030. The confrontation between the available water resources and the increasing demand for water shows that this demand will not be satisfied by 2030 (Fig. 24) without the implementation of more effective conservation and water saving strategies as well as water infrastructure development programs (MARH 2007). Projections of average monthly rainfall

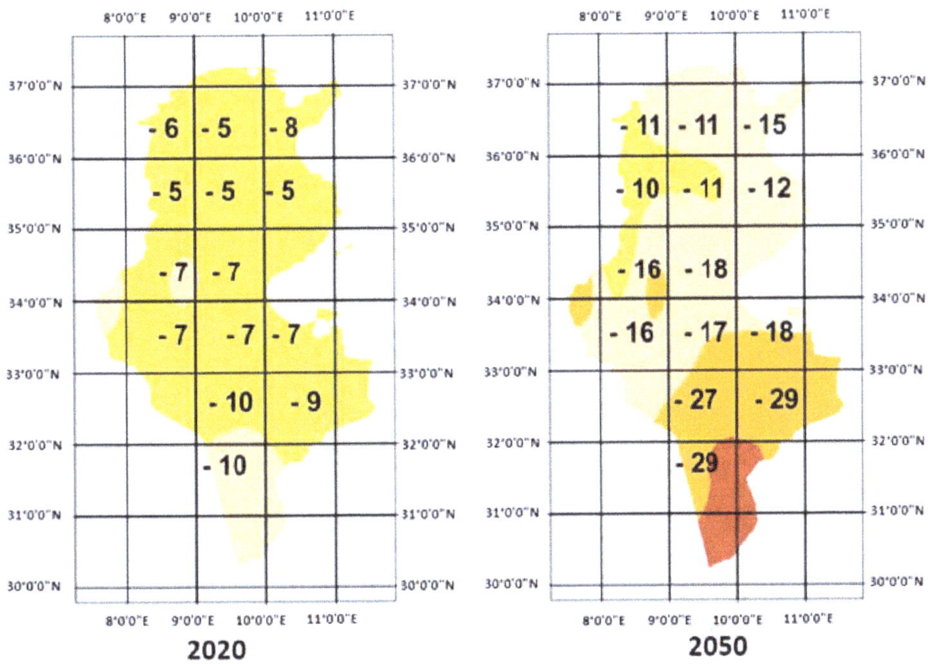

Fig. 22 Decreases of the average annual rainfall in % of the HadCM3 model (A2 scenario) compared to the reference period by 2020 and by 2050 (GTZ 2007)

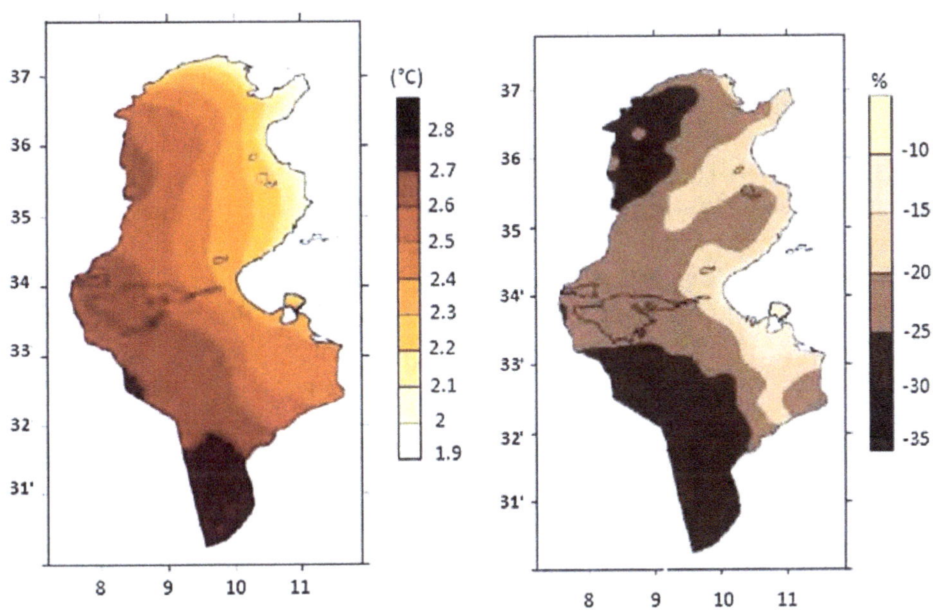

Fig. 23 Projection of temperature and rainfall by 2100 (INM in Ouasli 2015)

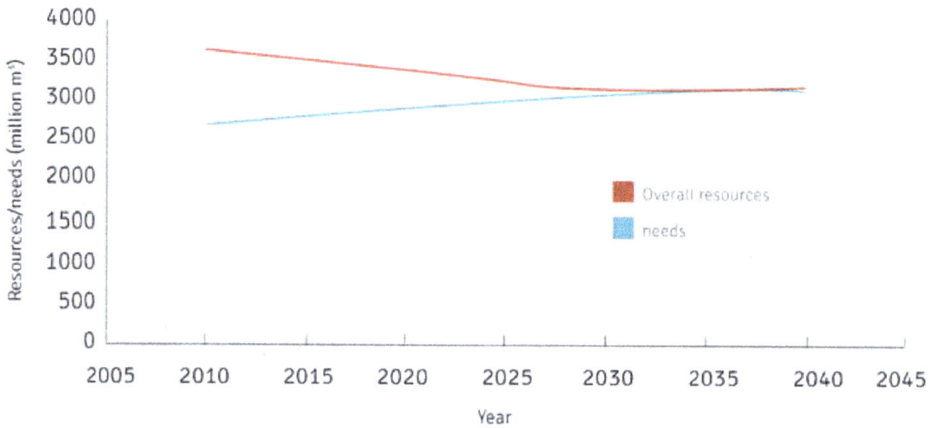

Fig. 24 Evolution (in million m^3) of overall water resources (in blue) and needs for water (in red) in Tunisia taking into account climate change impacts on water resources (MARH 2007)

over Tunisia show the future evolution of aridity and mark its trend to the north of Tunisia which shows the vulnerability of the south to water stress in the future (Ouasli 2015).

Recent climate projections confirm the reduction of piezometric levels of the aquifer in the Sfax region located in southeast of Tunisia by 2020 and 2050 for the following two scenarios (Boughariou et al. 2018): the first considers a constant and increasing consumption, the second integrates climate projections published by the Tunisian Ministry of Agriculture and Water Resources and the German Agency for International Cooperation (GIZ), using HadCM3 as a model for general circulation (Boughariou et al. 2018). Both scenarios also indicate the quantitative degradation of the groundwater by 2050 with an alarming marine intrusion (Boughariou et al. 2018).

The projection of water resources, on the basis of various hypotheses proposed, shows that the exploitable conventional water resources will be around 3,829 million m^3 in 2030, given that the water management program is carried out entirely (high hypothesis), of 3170 Mm^3 in the case where only a part of the program will be realized (average hypothesis) and 2718 Mm^3 and in the case where the program will not be realized (hypothesis low) (TSNCUNFCCC 2013).

The addition of unconventional water resources will strengthen the exploitable resources. This is the desalination of seawater (80 Mm^3 in the year 2030) and the reuse of treated wastewater (292 Mm^3) for agricultural needs. The total exploitable in Tunisia in 2030 will be around 4,201 Mm^3 (TSNCUNFCCC 2013).

Total water requirements by 2030 are projected at about 2911 Mm^3 in the case of the low assumption, 3,054 Mm^3 for the medium hypothesis and 3,743 Mm^3 for the high hypothesis. Needs-side assumptions are also defined based on the rate of achievement of the projected programs (TSNCUNFCCC 2013).

In the case where all programs for water mobilization are carried out, the projected needs can be met by 2030 (an excess of about 11% of the available amount is planned, Table 2). In the case where the planned programs are not all realized, a

Table 2 Evolutions of water resources considering climate change (in millions of m^3)

Designation	2010			2020			2030		
	Potential	Mobilized	Exploitable	Potential	Mobilized	Exploitable	Potential	Mobilized	Exploitable
Dams	2700	2121	1378	2700	2131	1385	2700	1890	1229
Groundwater	758	758	758	781	781	591	805	805	308
Deep aquifers	1544	1350	1350	1791	1535	1215	2079	1731	1214
Total conventional waters	5002	4229	3486	5272	2447	3191	5584	4426	2751
Treated wastewater	253	117	117	400	203	203	512	372	372
Desalted waters		18	18	0	47	47	0	80	80
Total non-conventional waters	253	117	117	400	203	203	512	372	372
Total	5255	4336	3603	5672	4650	3394	4798	4798	3123

Source: Ministry of Agriculture and Water Resources (MARH 2007)

lack of water is likely to appear from 2022 (and this, even if the low hypothesis of the needs side is retained) (TSNCUNFCCC 2013).

If water needs do not evolve as planned and programs to increase the availability of the resource are not realized (low hypothesis), the difference between needs and available water resources would be negative by 2015 (TSNCUNFCCC 2013).

The vulnerability of conventional water resources to climate change in Tunisia is characterized by (Ouasli 2015):

- A reduction of these resources by around 28% by 2030. The decrease will affect groundwater, coastal water, and nonrenewable groundwater.
- Increasing frequencies of extreme events: drought and flood.
- The reduction of water volumes in dams by around 5% by 2030.
- A deterioration of the quality of surface and underground waters.
- The succession of extreme events: floods and drought.
- The succession of periods of drought. These periods will have an impact on the natural recharge of groundwater. These aquifers will be exposed to over-exploitation to compensate the deficit in surface water for irrigation.

Adaptation

Water Saving Measures

The main threats to water resources are:

- The overuse of groundwater resources
- Pollution of water resources (rivers, groundwater, etc.). Discharges by solids or liquids into the environment, rivers, and vulnerable groundwater
- The silting of reservoirs
- The incidence of successive dry years that hinder the annual renewal of water resources
- Water wastage and losses
- The impact of climate change

It was recorded around 20 droughts during the twentieth century. During this period a severe drought is recorded every 6 years.

During the years 1987, 1988, and 1989, Tunisia has experienced the most critical drought. It led to a water deficit around 30%.

In case of drought, the agricultural sector, which currently consumes 81% of water resources, will suffer water shortages which will result in an increased use of groundwater and can have dramatic effects, since these resources contribute to maximum of 82% to the satisfaction of agricultural needs (Fig. 25).

To overcome the lack of water in the future, the below directions are to be followed:

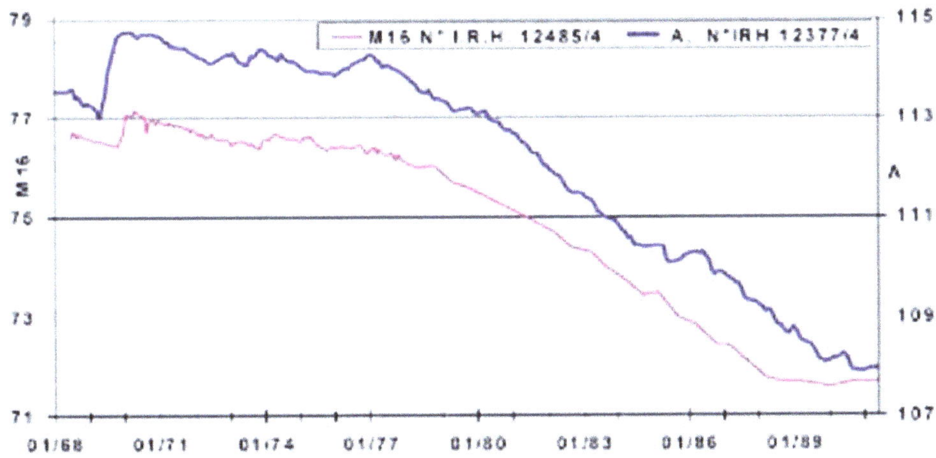

Fig. 25 Decrease of the level of aquifer Merguellil (central Tunisia)

- Complete the implementation of works to mobilize water resources to achieve a mobilization rate of 95%
- Integrated management of water resources by capitalizing excess of wet years in order to reduce the effects of drought
- The water conservation and demand management in all sectors
- The use of wastewater in agriculture
- The use of desalination water for drinking water
- The protection of water resources against pollution
- The protection of groundwater against overexploitation (Fig. 25)

The presidential program "Let us rise to the challenges 2009–2014" provides for an increase of 30% currently to 50% in 2014 the exploitation rate of treated wastewater in agriculture, industry, and tourism.

Large areas of around 8,500 ha of irrigated crops with treated wastewater will be built within this framework, in the Greater Tunis, Zaghouan, Beja, Sousse, Kairouan, Sfax, Gafsa, Kébili, Gabes, and Tataouine Medenine.

Meanwhile, the authorities will consider the study of potentialities for food, with unconventional resources, groundwater, notably those excessively exploited.

The desalination of water in the majority belongs to the SONEDE for a total capacity of 95,000 m^3 (Lahacher and Pillet 2008). Desalination of water is being expanded especially in the tourism sector. New desalination installations based on sea water are programmed by SONEDE in Djerba, Sfax, and Zarrat for a total potential of about 80 m^3 in 2030.

Recommendations

The following proposed measures may be the best way to cover future water demand (Ouasi 2015; Ben Nouna et al. 2018):

- Improve the consideration of ecosystems: necessary to ensure adequate recharge of aquifers for the production of good quality of water
- Improvement of water demand management and strengthening of the National Water Saving Program especially water saving for irrigation
- Develop a strategic policy regarding the prediction of extreme events, raising of dams to safeguard their capacity, encouraging the practice of less water-consuming crops
- Quantify the processes of water erosion and siltation rates of dams
- Control of energy consumption during the production of water (pumping, desalination) using clean energies (solar for example)
- Develop integrated management of water resources by basin
- Involve farmers so that they can participate in the future of their water resources
- Enhance knowledge and experience in the field of traditional hydraulic systems
- To further develop the artificial recharge of groundwater
- Development of scientific research related to the water sector

Conclusion

According to recent studies, the Maghreb countries will become hotspots by the end of the twenty-first century. Higher temperatures and reduced rainfall will increase the frequency of droughts in these countries. The drought observed in recent decades has tragic consequences. These effects are reflected by lower inflows to dams with more than 30% reduction in dam volumes and the lower level of groundwater sometimes with 40 m of drawdown in the plain of Mitidja in Algeria for example. This latter is the combined effect of drought and overexploitation of groundwater. This overexploitation is due to the decrease of the natural recharge and the increase of the water demand (Increasing demography and economic and social development).

The decrease in rainfall, as the lag time of rainy seasons, has penalized development projects particularly related to agriculture and affect the proper functioning of the carried out infrastructure.

To mitigate the impacts of climate change in the region, we recommend the integration of climate change in the programs and policies particularly in the context of sustainable development. It must also take into account the weather, and reinforce the measures of Weather assistance in the rural areas and to all users.

The undeniable effects of drought on reducing the level of production of cereals in particular require from planners to take into account the continuing risks associated with an arid climate and the risks of seasonal drought in all aspects of agricultural development. Thus the long dry season causes prolonged water stress to be taken into account in the selection of appropriate crops and varieties.

For the three countries, many projects and programs are developed and implemented to cope with the effects of climate change and Anthropic. These countries are also facing rapid population growth and sustained socioeconomic development. Agriculture is the cornerstone of this development, which will increase the demand for water. The water infrastructure programs must take into account this strategy which will ensure the food security of these three Maghreb countries.

In order to meet this ever increasing demand for water, especially for agriculture, it is necessary to set up special programs and develop irrigation techniques to reduce the consumption of water resources in agriculture. These programs should also make it possible to increase agricultural production to ensure food security in these three countries. Also, the reduction of water leakage from drinking water supply networks that are important in the order of 30 to 40%. The fight against water scourge makes it possible to create new irrigated perimeters and to effectively supply the population with drinking water.

References

Agoumi A (2003) Vulnérabilité des pays du Maghreb face aux changements climatiques. International Institute for Sustainable Development (IISD), Winnipeg. 11 p

Alaoui M (2013) Water sector in Morocco: situation and perspectives. J Water Resour Ocean Sci 2 (5):108–114

Ameur F (2017) Construction de la surexploitation et reproduction des inégalités d'accès et d'usage des eaux souterraines. Thèse de doctorat. AgroParisTech, France.

Amraoui F (2019) Integrated management of water resources as a response to the water deficit: Case of the Saiss Plain (Morocco). International conference "groundwater resilience to climate change and high pressure" GRECPIMA 2019. May -8-10, 2019, Muğla, Turkey

Bct (2007) Rapport annuel de la banque centrale de Tunisie. Ministère du Développement et de la Coopération Internationale de Tunisie. 70 p. https://www.apbt.org.tn/wp-content/uploads/2016/06/rapport_fran_2007_t1.pdf

Ben Abdelfadel A, Driouech F (2008) Climate change and its impacts on water resources in the Maghreb region. 13èmeCongrèsMondial de l'Eau, 1–4 septembre, Montpellier, France

Ben Nouna B, Rezig KS, Chebil A, Jebari S (2018) Water resources and climate change: key adaptation strategies and the role of scientific research in Tunisia. J New Sci Agricult Biotechnol 51(1):3138–33147

Bougharious E, Allouche N, Jmal I, Mokadem N, Ayed B, Hajji S, Khanfir H, Bouri S (2018) Modeling aquifer behaviour under climate change and high consumption: Case study of the Sfax region, southeast Tunisia. J Afr Earth Sci 141:118–129. https://doi.org/10.1016/j.jafrearsci.2018.02.006

Chebil A, Souissi A, Frija A, Stambouli T (2019) Estimation of the economic loss due to irrigation water use inefficiency in Tunisia. Environ Sci Pollut Res 26:11261–11268. https://doi.org/10.1007/s11356-019-04566-8

Choukr-Allah R. Comparative study between Moroccan water strategies and WFD. In: Junier S, El Moujabber M, Trisorio-Liuzzi G, Tigrek S, Serneguet M, Choukr-Allah R, Shatanawi M, Rodríguez R (eds) Dialogues on Mediterranean water challenges: Rational water use, water price versus value and lessons learned from the European Water Framework Directive. Bari : CIHEAM, 2011. p. 181–188 (Options Méditerranéennes : Série A. Séminaires Méditerranéens; n. 98)

Dahan S (2017) Managing urban water scarcity in Morocco. World Bank, Washington, DC

Demmak A (2008) La sècheresse en Algérie des années 1975/2000. Second meda water regional event on local water management, Marrakech, morocco, 28, 29 & 30 April 2008

Driouech F (2010) Distribution des précipitations hivernales sur le Maroc dans le cadre d'un changement climatique: Descente d'échelle et incertitudes. PhD thesis University of Toulouse (France), 163 p

Droogers P, Immerzeel W W, Terink W, Hoogeveen J, Bierkens MFP, Van Beek LPH, Debele B (2012). Water resources trends in Middle East and North Africa towards 2050. Hydrol Earth Syst Sci, 16, 3101–3114., http://www.hydrol-earth-syst-sci.net/16/3101/2012/. https://doi.org/10.5194/hess-16-3101-2012

El Hafid D, Zzerrouqi Z, Akdim B (2017) Study of drought sequences in the Isly basin (East Morocco). Larhyss J 31:83–94

Elloumi F (2016) La gouvernance des eaux souterraines en Tunisie. IWMI projectreport n°7. Groundwater governance in the Arab World. 2016. 121 p

Farzaneh MR, Eslamian S, Mirnezami SJE (2014) Chap 8 Climate change: uncertainty, impact, and adaptation. In: Eslamian S (ed) Handbook of engineering hydrology, Vol. 2: modeling, climate changes and variability. Francis and Taylor, CRC Group, Boca Raton, pp 127–146

Giorgi F (2002) Variability and trends of sub-continental scale surface climate in the twentieth century. Part I: observations. Clim Dyn 18:675–691. https://doi.org/10.1007/s00382-001-0204-x

GTZ (2007) Stratégie Nationale d'Adaptation de l'Agriculture Tunisienne et des écosystèmes aux changements climatiques. Deutsche Gesellschaft für Internationale Zusammenarbeit (GIZ) GmbH. 150 p. http://www.environnement.gov.tn/PICC/wp-content/uploads/Strat%C3%A9gie-nationale-d%E2%80%99adaptation-de-l%E2%80%99agriculture.pdf

Habibi B, Meddi M, Torfs PJJF, Remaoun M, Van Lanen HAJ (2018) Characterisation and prediction of meteorological drought using stochastic models in the semi-arid Chéliff–Zahrez basin (Algeria). J Hydrol Reg Stud 16:15–31. https://doi.org/10.1016/j.ejrh.2018.02.005G

Haddour A, Mahé G, Meddi M (2019) Impact des changements climatiques sur la ressource en eau dans trois grands bassins versants du centre et de l'ouest de l'Algérie. Séminaire international sur l'hydrogéologie et l'environnement. 12–14 mars 2019, Ouargla (Algérie)

Hallouz F, Meddi M, Mahe G (2013) Modification du régime hydroclimatique dans le bassin de l'oued mina (nord-ouest d'Algérie) – changes of the hydroclimaticregime in the basin of the Wadi mina (northwestAlgeria). Revue des Sciences de l'Eau 26(1):33–38. https://doi.org/10.7202/1014917ar

Hamza M (2007) Gestion des ressources en eau en Tunisie. ineco- Nabeul, le 6 décembre 2007. http://environ.chemeng.ntua.gr/ineco/UserFiles/File/TunisiaWorkshop/Gestion%20des%20ressources%20en%20eau%20en%20Tunisie.pdf

INDC (2015) Morocco's intended nationally determined contribution, 36 p. https://www4.unfccc.int/sites/submissions/INDC/Published%20Documents/Morocco/1/Morocco%20INDC%20submitted%20to%20UNFCCC%20-%205%20june%202015.pdf

IPCC (2013) Summary for policymakers. In: Stocker TF, Qin D, Plattner G-K, Tignor M, Allen SK, Boschung J, Nauels A, Xia Y, Bex V, Midgley PM (eds) Climate change 2013: the physical science basis. Contribution of working group I to the fifth assessment report of the intergovernmental panel on climate change. Cambridge University Press, Cambridge/New York, pp 3–30

Jacobeit J (2000) Rezente Klimaentwicklung im Mittelmeerraum. Petermanns Geogr Mitt 144:22–33

Kahlerras M, Meddi M, Benabdelmalek M, Tumi S, Kahlerras D, Nouiri I (2018) Modeling water supply and demand for effective water management allocation in Mazafran basin (north of Algeria). Arab J Geosci 11(18):1–12

Khomsi K, Mahe G, Sinan M, Snoussi M (2013) Hydro-climatic variability in two Moroccan watersheds: a comparative analysis of temperature, rain and flow regimes. In: Boegh E, Blyth E, Hannah DM, Hisdal H, Kunstmann H, Su B, Yilmaz KK (eds) Climate and land surface changes in hydrology, vol 359. IAHS Publications, Wallingford, pp 183–190

Knippertz P, Christoph M, Speth P (2003) Long-term precipitation variability in Morocco and the link to the large-scale circulation in recent and future climates. Meteorog Atmos Phys 83:67–88. https://doi.org/10.1007/s00703-002-0561-y

Lahacher RG, Pillet G (2008) Projection des ressources en eau de la Tunisie sous les effets des changements climatiques globaux. XIIIth World Water Congress, 1–4 septembre, Montpellier, France

MARH (2007) (Ministère de l'Agriculture et des Rein climate change 2013 ssources Hydrauliques de Tunisie). Stratégie Nationale d'Adaptation de l'Agriculture Tunisienne et des écosystèmes aux changements climatiques. Bureaux d'études GOPA et Exa Consult Tunisie, janvier 2007. 150 p

Matari A (2016) Climat et changement climatique. Journal Algérien des Régions Arides (jara), Numéro Special, 2016

MDCE (2016) Troisième communication Nationale du Maroc à la convention cadre des nations unies sur le changement climatique. Ministère Délègue Auprès du Ministère de l'Énergie, des Mines, de l'Eau et de l'Environnement Chargé de l'Environnement, Avril 2016. https://unfccc.int/resource/docs/natc/marnc3.pdf

Meddi M, Boucefiane A (2009) Impact of Climate Change on the renewal of underground water resources (Case Hydrographic Basin Chéliff-Zahrez). Groundwater and Climate in Africa (Proceedings of the Kampala Conference, June 2008). IAHS Publication, 334

Meddi M, Hubert P (2003) Impact de la modification du régime pluviométrique sur les ressources en eau du Nord-ouest de l'Algérie. In: Servat E, et al., eds. Hydrology of Mediterranean and semi-arid regions. Wallingford: IAHS Press, IAHS. Publ. 278, 229–235. Disponible á: http://iahs.info/redbooks/a278/iahs_278_229.pdf

Merabti M, Martins DM, Meddi M, Pereira lS (2018) Spatial and time variability of drought based on SPI and RDI with various time scales. Water Resour Manag 32:1087–1100. https://doi.org/10.1007/s11269-017-1856-6

Mitchell TD, Hulme M, New M (2002) Climate data for political areas. Area 34:109–112. http://www.cru.uea.ac.uk/timm/data/index-table.html

New M, Todd M, Hulme M, Jones P (2001) Precipitation measurements and tends in the twentieth century. Int J Climatol 21:1899–1922

Nouaceur Z (2016).Analyse de la variabilite pluviometrique sur près d'un demi-siècle au Maghreb central (Maroc, Algérie, Tunisie). XXIXe Colloque de l'Association Internationale de Climatologie, Lausanne – Besançon

OSS (2009) Vers un système d'alerte précoce à la sécheresse dans 3 pays de la rive sud de la Méditerranée. Collection de Synthèse n°4. OSS – Tunis 2009. 86 p. http://www.oss-online.org/cd_envi/doc/07/03/06.pdf

Ouasli A (2015) Secteur de l'eau en face des changements climatiques (cc). Forum scientifique "innovation dans le secteur agricole et le défi du changement climatique". Bureau de Planification et des Equilibres Hydrauliques. République Tunisienne, Ministère de l'Agriculture, des Ressources Hydrauliques et de la Pêche. http://www.siat.tn/medias/files/siat-2016/interventions/2-2-Secteur-de-l-Eau.pdf

Rodrigo FS, Trigo RM (2007) Trends in daily rainfall the Iberian Peninsula from 1951 to 2002. Int J Climatol 27:513–529

Sarra A (2009) Projections, impacts et stratégies d'adaptation. https://www.webmanagercenter.com/2009/08/26/79540/tunisie-changement-climatique-projections-impacts-et-strategies-d-adaptation-3-3/

SCNCCRM (2009) Seconde Communication Nationale à la Convention Cadre des Nations Unies sur les changements climatiques du royaume du Maroc. Ministère de l'Aménagement du Territoire, de l'Urbanisme, de l'Habitat et de l'environnement, 2009. 217 p

SIGMA (2017) L'eau en Tunisie entre enjeux stratégiques et qualité de vie des citoyens. Les rtencontres de Tunis, Conférence – débat. 28 septembre 2017

Sinan M, Belhouji A (2016) Impact of the climate change on the climate and the water resources of Morocco on horizons 2020, 2050 and 2080 and measures of adaptation. La Houille Blanche 4:32–39

Sinan M, Boussetta M, El Rherari A (2009). Changement climatique: causes et conséquences sur le climat et les ressources en eau. J HTE 142 21–30

Taabni M, El Jihad MD (2015) Eau et changement climatique au Maghreb: quelles stratégies d'adaptation? Les Cahiers d'outre-Mer 65(260):493–518

Trabelsi F (2016) Impact du changement climatique sur les ressources en eau souterraine et politiques d'adaptation en Tunisie. Project: Tempus IV: Building Climate Change Adaptation in Morocco, Algeria and Tunisia "Climadapt"

Tramblay Y, El Adlouni S, Servat E (2013) Trends and variability in extreme precipitation indices over Maghreb countries. Nat Hazards Earth Syst Sci 13:3235–3248. https://doi.org/10.5194/nhess-13-3235-2013

TSNCUNFCCC (2013) Tunisia's second national communication under the United Nations framework convention on climate change (UNFCCC). 174 p., 2013. http://www.unfccc.int/resource/docs/natc/tunnc2.pdf

World Bank (2013) La Tunisie face aux changements climatiques. Évaluation et Actions pour accroitre la résilience et le développement. 2010. 174 p

World Bank (2017) Beyond scarcity: Water security in the Middle East and North Africa. MENA development report. World Bank, Washington, DC. Licence: CC BY 3.0 IGO https://openknowledge.worldbank.org/handle/10986/27659

World Bank Group (2017a) Climate variability, drought, and drought management in Morocco's agricultural sector in Morocco. http://documents.worldbank.org/curated/en/353801538414553978/Climate-Variability-Drought-and-Drought-Management-in-Moroccos-Agricultural-Sector

World Bank Group (2017b) Managing urban water scarcity in Morocco. World Bank, Washington, DC. © World Bank. https://openknowledge.worldbank.org/handle/10986/29190. License: CC BY 3.0 IGO

Zeroual A (2017) Evolution du climat passé récent et futur en algérie. PhD thesis, ENSH, Blida, 2017, 250 p

Zeroual A, Assani Ali A, Meddi M (2017) Combined analysis of temperature and rainfall variability as they relate to climate indices in northern Algeria over the 1972–2013 period. Hydrol Res 48(2):584–595

4

Climate Change Adaptation and Community Development in Port Harcourt, Nigeria

Julie Greenwalt, Michael Dede, Ibinabo Johnson, Prince Nosa, Abi Precious and Barbara Summers

Contents

J. Greenwalt (✉)
Go Green for Climate, Amsterdam, Netherlands
e-mail: julie@gogreenforclimate.com

M. Dede · I. Johnson · P. Nosa · A. Precious
Chicoco Collective, Port Harcourt, Nigeria

B. Summers
CMAP, Port Harcourt, Nigeria
e-mail: barbara@cmapping.net

Abstract

Port Harcourt, Nigeria, as with many cities in Africa, is already experiencing impacts from climate change while also facing development challenges and compounding vulnerabilities. There is often a gap between the adaptation recommendations of academia and international organizations and the conditions needed to achieve these recommendations given the realities on the ground in cities with vulnerable populations, pressing challenges, and minimal capacity. This gap can make it difficult to translate theoretically persuasive plans and guidance into practically sustainable actions. This chapter builds on the experience of CMAP, a Port Harcourt-based NGO leading participatory mapping, community media, and public space design programs in the city's informal waterfront settlements. Drawing on the perspectives of Chicoco Collective, the youth-led volunteer network which CMAP supports, and the datasets that they have built, the chapter explores local awareness and skills for climate change adaptation. This chapter concludes with recommendations relevant for Port Harcourt and similar African cities to build on such local skills and experiences and advocates for a partnership-based approach that brings together adaptation professionals and community-based actors for more effective critical analysis of local conditions and prioritization of actions to meet the development and climate change needs of local communities.

Keywords

Community development · Port Harcourt · Nigeria · Flooding · Community data · Awareness raising · Informal settlements

Introduction

There is growing global awareness and urgency about the need to accelerate action to adapt to climate change. Many cities and local communities, especially in Africa, are already experiencing the impacts of climate change, and risks and vulnerabilities are increasing. This chapter focuses on Port Harcourt, Nigeria, which is a secondary city in the most populous country in Africa and the oil capital of Nigeria. The chapter will share data and information gathered by CMAP, a local NGO, and Chicoco Collective, the youth-led volunteer network that it supports, to explore the risks and vulnerabilities in Port Harcourt, as well as the ongoing work of CMAP and the perspective of local communities, to identify some of the key challenges and to propose recommendations for how Port Harcourt and similar African cities could accelerate action on climate change.

The chapter also explores the gap between the adaptation recommendations of academia and international organizations and the conditions needed to achieve these recommendations given the realities on the ground in cities with vulnerable populations, pressing challenges, and minimal capacity. This gap can make it difficult to translate theoretically persuasive plans and guidance into practically sustainable actions. However, a partnership-based approach that brings together adaptation professionals and community-based actors for more effective critical analysis of local conditions and prioritization of actions can help to advance both the development and climate change adaptation needs of local communities.

This chapter presents findings from focus groups, mapping, surveys, and transect walks undertaken by CMAP and Chicoco Collective related to community development, vulnerability, and climate change awareness. It also provides a brief introduction to Port Harcourt, climate change in Nigeria, and a brief review of some of the latest literature and recommendations produced at the global level on adaptation to climate change in cities, Africa, and informal settlements.

Port Harcourt, Nigeria

Port Harcourt is the capital of Rivers State and is located in the Niger Delta region. The city lies at the mouth of the River Bonny. It is located approximately 25 km from the Atlantic Ocean and is situated between the Dockyard creek/Bonny River and the Amadi creek. It lies at an average altitude of about 12 m above mean sea level, although the waterfront communities tend to be below 4 m (Fig. 1).

Fig. 1 Map of the Niger Delta highlighting Port Harcourt's location 25 km inland from the Atlantic Ocean (created by CMAP in March 2020 using QGIS and data from Rivers State Office of the Surveyor General)

Port Harcourt was founded in 1912 during the colonial era as a railroad terminus and deep-water port for exporting coal, then later, oil and gas, after their discovery in the late 1950s in the Niger Delta. The oil and gas sector continues to dominate the local formal economy. However, despite the economic wealth generated in the region, many people in the city and state live in impoverished conditions. IMF-imposed structural adjustment in the 1980s and 1990s saw Nigerian cities such as Port Harcourt simultaneously grow and physically deteriorate. Even as they absorbed newcomers, they often lacked the resources to provide appropriate infrastructure and services to growing populations. With an economy so closely connected to resource extraction, long term plans are vulnerable to disruption by oil and gas-market downturns, as well as changes in state government. Historically, a drop in oil prices has corresponded to a shortage of resources for housing, water supply, security, or waste management in Port Harcourt and is often linked with the proliferation of informal settlements (Bloch et al. 2015). A series of abandoned Port Harcourt master plans and infrastructure projects, most notably the monorail, which starts and ends abruptly and has never been used, are testament to the absence of sustained urban planning and action as a result of low economic and political investment.

Despite its current ranking as one of the largest economies in Africa, Nigeria has failed to translate growth in output into an increase in formal employment, and this failure has resulted in significant underemployment. This is an even greater challenge for the country's growing share of youth. Fifty-five percent of the population is under the age of 20 (World Bank 2016). This nationwide trend is evidenced in employment rates in Port Harcourt where not only has youth unemployment increased, but is coupled with significant underemployment and a lack of productive opportunities for young workers. For many youth, participation in the illegal/informal crude oil refining process in the creeks around the city is seen as one of the few – but most lucrative – options in the local economy. Further exacerbating the issues, the city once known as the Garden City is now characterized by degradation of waterways, soil, and air largely due to the oil and gas industry (both legal and illegal).

Nigeria is a federal state with much of the governance power resting in state governments which have ministries for environment, urban development, education, etc. Despite this decentralized governance system, active city governments are largely nonexistent, and state governments have primacy over physical development in urban areas. This is the case in Port Harcourt.

The city of Port Harcourt has an estimated population of over 1.8 million, making it the sixth largest city in Nigeria and the largest in the South-South region (OECD/SWAC 2020). The climate is tropical monsoon with an annual mean maximum temperature 30.9 °C and mean minimum temperature 22.6 °C.

As noted in the 2009 Max Lock Centre study by Theis et al., spontaneous and uncontrolled physical growth driven by rapid urbanization in Port Harcourt over the last several decades has increased the settlement of the areas alongside swamps in the waterfront. Faced with high-cost inner-city rentals and a shortage of housing, many new urban dwellers are forced to settle in high-risk neighborhoods exposed to flooding and the adverse effects of climate change. These waterfront communities are now the largest informal settlements in the city, with some population estimates of over 480,000 people (Theis et al. 2009). While they have varying levels of development, with little to no government intervention or service provision, they can be characterized as lacking in many basic services including sanitation, water, and waste disposal. These communities also regularly face the threat of forced eviction with several large-scale demolitions taking place in 2009, 2012, and 2016.

Communities First: Effective Community-Led and Data-Driven Change

Collaborative Media Advocacy Platform (CMAP) has been working with residents of informal waterfront communities in Port Harcourt for over 10 years. CMAP began as an informal association motivated by building both local and global awareness of, and halting, massive forced evictions in waterfront communities. The project's animating impulse was the response of community members to the camera of the organization's now-project director when he found himself in the middle of a large-scale forced eviction. In the middle of the demolition, residents rallied around the camera, protecting it from the police and directing what it filmed.

In 2009, the then Governor of Rivers State announced the planned demolition of all the city's waterfront settlements. These settlements were not represented on municipal maps. They featured in the city's development plan only as undifferentiated zones marked for "clearance" through "major relocations." The resultant forced evictions and demolitions triggered not only widespread displacement, but also a conflict that threatened to become militant. In response, a group of local activists and international campaign organizations started to support mobilization of these marginalized waterfront communities, making visible the violence and disregard of state agencies through documentary photography, film, and recorded testimonies.

Following initial campaigning and legal advocacy work, there was a systematic broadening of engagement with different forms of representation – from the cinematic to the cartographic, from the voice on the street to the voice on the airwaves, and from architectural design to legal representation – and a strategic exploration of how they could be layered to support the struggle of Port Harcourt's waterfront settlements to be seen and heard. After several years of ad hoc activities, programs were solidified into the Human City Project, led by CMAP and a local community-registered trust. The project's programs – Chicoco Cinema, Chicoco Radio, Chicoco Sound, Chicoco Maps, and soon Chicoco Solar – are the focus of the Chicoco Collective. In its over 9 years of work, the project has established highly trained community mapping and media teams, including reporters, writers, presenters, broadcasters, filmmakers, photographers, actors, musicians, surveyors, and technical producers, with the capacity to manage a community radio and music studio.

The Chicoco Maps program, which launched as a direct response to communities not being recognized on any map or in official statistics, focuses on building the

Fig. 2 Household unit map of Ibiapu Polo Waterfront Community (CMAP 2018)

capacity of young waterfront residents to put their communities on the map, capture the information to tell their story on their terms, and generate the necessary data to negotiate with government authorities. Over the past 6 years, the volunteer community cartographers have created base maps of six waterfront communities and, for three of those communities, have developed the most thorough dataset on any community in the greater Port Harcourt area. An example of one of the community maps can be seen below which details structures and land use (Fig 2).

Their work has involved indexing every building, as well as conducting household surveys, gathering detailed community-level data on population, land use, employment, health, and service provision, among others. This data is fed into an encounter and exchange program in which civil society, government representatives, and community members are brought together in solution-focused discussions facilitated by project participants around their data and map outputs in each community.

Community Mapping and Data

The Chicoco Mapping team carried out detailed demographic and structure/infrastructure surveying of three waterfront communities. The team was able to survey

Table 1 Demographics of three waterfront communities in Port Harcourt: Ibiapu, Darick, and Amatari Polos (CMAP 2018)

	Ibiapu Polo	Darick Polo	Amatari Polo
Population	4,079	1,922	1,846
Number of households	969	447	504
Population density (ppl/km^2)	73,833	35,900	67,000
Number of businesses	126	69	97

77–80% of households in each community, which gave a representative number of responses to extrapolate total population as well as to understand the major trends in the community. The following overview statistics of Ibiapu Polo, Darick Polo, and Amatari Polo communities can be seen in Table 1.

These communities share similar topographical characteristics with other waterfront communities throughout the city, being 10–15 m below the rest of the city on land reclaimed from mangroves. Amatari Polo is the oldest of the three, settled over 60 years ago. Ibiapu Polo is approximately 40 years old, while Darick Polo is less than 20 years old and is the only community that has been sand-filled. The others are built on "chicoco" mud, extracted from the mangroves. Despite the majority of residents having lived in their communities for 5 years or longer, households predominately reported having only informal or community-granted documentation of rent or ownership.

All three communities have a young population, with over 50% of residents under the age of 25. Unemployment rates were reported as 10–12%, but it is likely that actual unemployment rates are higher, with many people reporting self-employed and not employed full-time. These waterfront communities are all managed by an Okrikan system of community leadership led by an elected community chairman and supported by a range of committee members including a woman leader, youth leader, sanitation chairman, light chairman, etc.

While the community mapping project was initially focused on collecting data for preventing forced evictions or negotiating with the government on security of tenure or service provision, it is a useful basis for understanding some of the existing climate change impacts or highlighting potential vulnerabilities. For instance, the household surveys found that over half of households reported that flooding after a rainfall event restricts their movement. Furthermore, 29% of households reported that they always have to wait a while after it rains before they can leave their home. The chart below highlights the impacts of flooding on individual households (Fig. 3).

Infrastructure networks including roads and pathways, electrical, water, and waste networks have also been documented in detail in the mapping process. This is often the first time community scale infrastructures have been documented. This aggregation of infrastructure often highlights the ad hoc manner in which drainage is implemented in the community and between bordering communities when the various community data are collated. For example, the map below indicates that the drainage network in Ibiapu Polo, one of the neighborhoods intensively mapped and surveyed by the Chicoco team, is not integrated and few of its channels are

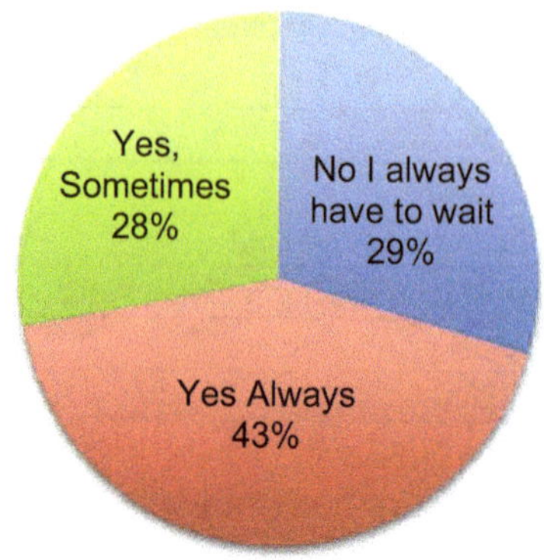

Fig. 3 Households in Ibiapu, Darick, and Amatari Polo reporting the ability to leave their home after a rainfall event. The response options were that yes, they can always leave their home; yes, they sometimes have to wait; or no, they always have to wait for flooding to recede (CMAP 2018)

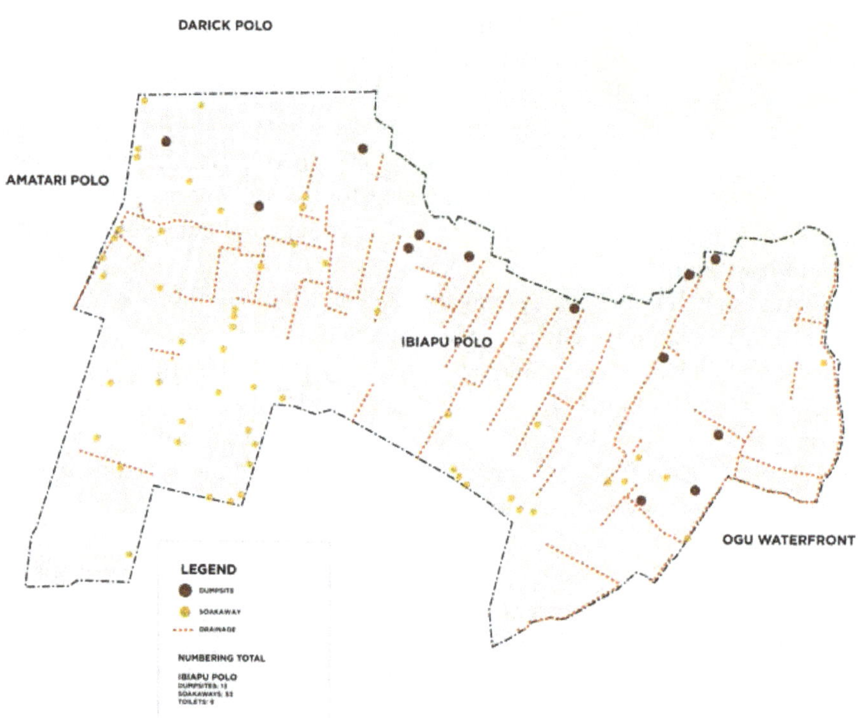

Fig. 4 Waste network map of Ibiapu Polo Waterfront Community, CMAP (2018)

ultimately connected to the larger network. These maps can serve as a base for understanding flooding bottlenecks and planning for improvements in each community. Insights into impacts of the poor drainage system and flooding are deepened by the household survey. Between 60% and 78% of households reported that drainage was one of the major problems facing their household (Fig. 4).

Another major problem raised by residents in the household survey was access to drinking water because the majority of households reported needing more than one source of drinking water. The two most significant sources of water reported are sachet water (i.e., water sold in small, sealed plastic bags) and private boreholes/taps outside their house. Less than one third of all households surveyed had access to piped water within their home.

Other issues raised by households in the survey included sewage leakage/lack of toilet facilities and management of solid waste. While these issues may not directly link to impacts of climate change at this time, they are likely to continue to exacerbate the impacts of climate change in these already vulnerable communities.

Documenting Sanitation and Exacerbating Conditions

There have been impacts at the community scale of the mapping work, but more significantly a shift in the citywide development dialogue. An African Development Bank-funded state water and sanitation services project in Port Harcourt claimed to be planning to reach 1.5 million people in Port Harcourt with improved WASH services. Yet, no waterfront communities or informal settlements were included in the initial designs. Data gathered by the Chicoco Maps team demonstrated the impossibility of reaching that many people unless the informal spaces of the city were included. In response, the project design was revised to include the city's waterfront settlements. Subsequently, as part of this larger project, in 2017, the mapping team led the data collection component of the World Bank-funded study "Technical Assistance to Fecal Sludge Management (FSM) Services in Port Harcourt, Nigeria."

While the data collection of this FSM study covered a range of urban morphologies, socio-economic conditions, and geographic locations, the project included a focus on low-income communities in Port Harcourt, and therefore a significant focus on waterfront communities. While the attention was primarily on documenting sanitation conditions and service provision, a number of climate-change-related risks and coping mechanisms were documented throughout the data collection process.

Both quantitative and qualitative research methods were employed to explore FSM practices along the entire service chain of collection, transport, treatment, and safe end use or disposal (World Bank 2017a). One of the data collection instruments used was a transect walk, during which enumerators walked through a community with several local residents to identify risks to public health, particularly with regard to the presence of fecal material and solid waste, and the proximity of these wastes to drainage channels and water resources. The World Bank defined objective was to observe environmental and public health risks along specified routes throughout the community.

The tool was revised by the community mapping team to fit the local context with revised survey questions and local terminology and piloted in two waterfront communities. The transect walk process was designed so that information on

Fig. 5 A flash flood in Port Harcourt Old Town during rainy season (photo credit CMAP 2016). (Permission obtained)

community characteristics, observed risks, and an estimate of how frequently these risks occur could be documented for each route and supplemented with GPS points and photo documentation. Observed risks were scored on a scale of "1 to 5," with a score of "5" denoting a particularly high risk.

Moreover, transect walks were helpful in: identifying and explaining the cause and effect of relationships among water uses, wastewater treatment, and sanitation conditions; identifying major problems and possibilities perceived by community members in relation to features or areas along the transect; providing an understanding of local technology and practices; and helping to triangulate and/or validate data collected through other tools.

Of the mix of 15 low-income informal settlements, middle-income neighborhoods, and high-income gated communities documented through transect walks, 73% experienced persistent flooding as a result of heavy rainfall and/or high tide (World Bank 2017a) (Fig. 5).

Furthermore, high risks were found to be more prevalent in low-income communities than middle- or high-income communities. Observed risks in these areas included unofficial dumpsites containing both solid and fecal wastes, pit latrines emptying into open drains, and evidence of open defecation. Low-income communities received a "high risk" rating of "4" or "5" for 75.8% of the observed conditions. In comparison, middle- and high-income communities received a high-risk rating for only 17.3% of the observed conditions (World Bank 2017a) (Figs. 6 and 7).

Fig. 6 The Chicoco mapping team conducting transect walks with community members to observe and record sanitation conditions in Tonipriama Ama Waterfront (photo credit CMAP 2017). (Permission obtained)

Fig. 7 The Chicoco mapping team conducting transect walks with community members to observe and record sanitation conditions in Baptist Waterfront (photo credit CMAP 2017). (Permission obtained)

The community team also received significant training on facilitating focus group discussions and developed a comprehensive questionnaire to gather qualitative data to compliment, validate, and/or challenge responses reported in the household survey. Information obtained from these discussions provided insight into household and community practices, service levels, and perceptions of past efforts to improve FSM services.

A focus group discussion was conducted at all 15 transect walk sites. Some focus group discussions were gender-specific. This was done because in mixed-gender sessions, men frequently dominated the discussions. Some men-only sessions were also held. Findings from these community focus group discussions stressed flooding challenges in the communities and informal methods being used to cope. Respondents in several communities mentioned that they themselves or others in the community were using nets or palm fronds to try to keep the solid waste and other debris that comes with floodwaters out of their homes. In three of the waterfront communities, it was mentioned that the community constructed a barrier, in some cases a metal plate or berm made of solid waste, along the riverbank to prevent flooding from seasonal high tides (seen below in Fig. 8) (World Bank 2017a).

Through the discussions, it was further revealed that flooding issues in waterfront communities were from the river and its high tide, as well as from drainage issues from upland areas and in the community. One respondent emphasized that, during rainy season, nets and barriers are removed to allow for waste to flow into the river (seen below in Fig. 9). A number of people also mentioned it was common practice in their community for people to use sticks or brooms to push the stagnant water and waste back out to the river after a rainfall (Fig. 9).

Fig. 8 A solid waste berm being created along the waterfront in Afikpo waterfront community (photo credit CMAP 2017)

Fig. 9 Nets installed in Baptist waterfront community to protect from waste flowing back into the community (photo credit CMAP 2017)

These focus group responses on flooding and their associated sanitation risks are supported by the findings of the Shit/Excreta Flow Diagram (SFD) created for this FSM project, which highlighted that other than the 13% of fecal sludge that remains unemptied in soak pits, there is no existing treatment of fecal sludge in low income areas of Port Harcourt, and it is directly entering the environment (World Bank 2017b).

Data and Media Advocacy

One of the unique aspects of the Chicoco Collective is that it brings together mapping and data collection with a range of advocacy platforms. The community mapping team has worked to analyze their data and has developed presentations to help the community and broader stakeholders think about how to best use the information to improve their community. In Ibiapu Polo, for example, the team was able to tally up all of the daily household expenses on drinking water into a yearly cost for the community. This totaled over 18 million Naira (approximately USD 50,000). Working with the community leadership, the cohort developed a proposal that is now being used by the community. The proposal seeks support to bring together these resources at a community level to implement a solar-powered public borehole, which is estimated to cost around seven million Naira, to serve the entire neighborhood's potable water needs.

A group of community mappers and journalists have built pilot air quality monitoring devices to document air quality conditions throughout the city. While their homemade devices do not yet yield data of sufficient accuracy or compositional detail to support rigorous analysis, their data does evidence alarming air conditions and reveals telling temporal and spatial pollution patterns. A team of campaigning citizen scientists has been working on how to use the data to create community awareness of air quality issues and is developing a citywide campaign. They have been able to further support this work through audiovisual documentation of illegal oil refining activities and their impacts on air quality, as well as interviews with community members and government representatives, including the then Rivers State Commissioner for environment.

These themes have also come together powerfully in other cultural and communicational forms, including as issues addressed through an original radio drama series "Angala Community," set in a fictional waterfront community. The most recent season of the radio drama, which will be aired citywide on partner stations, is focused on the gendered dynamics of waterfront communities, with an emphasis on sanitation and environmental conditions, including air quality. Numerous radio programs have been written, recorded, and produced by the community mapping and media teams addressing environmental issues in waterfront communities. One program in particular investigated the causes of flooding in the community, exploring the consistent blockage of drains and creeks with solid waste and the construction of unapproved buildings on waterways. This is an issue worsening rapidly as the waterfronts densify and expand into the mangroves, further compromising the capacity of these ecosystems to maintain hydrological balance, further undermining local livelihoods, and exacerbating entrenched economic and environmental inequalities.

The Chicoco Collective deploys a range of creative research, outreach, and engagement methods to bring climate-change-related issues to the forefront of community concern and broader popular conversation. Chicoco Cinema has used an inflatable pop-up movie theatre to raise awareness and keep local communities in the picture, increasing the accessibility of the discourse. The Chicoco Band, which includes a number of mappers, is currently working on their own cover of the well-known Nigerian artist, Fela Kuti's, "Water No Get Enemy" to address the current and looming issues of access to potable water in their waterfront communities and the country. Music can be a powerful medium of movement-building and a critical way of translating research findings into adaptation action.

The Chicoco team has established a program for prioritizing focal areas for their work each year. Focal issues have included housing, gender dynamics, and most recently, climate change. These issues are discussed through weekly sessions: "Bringing It Home: local impacts of global issues." These sessions focus on cohort-led research and discussion to understand the global context of the issues and how they can be contextualized in Nigeria and in their communities. CMAP has supported these research efforts by connecting participants with area specialists and staging expert-led workshops and trainings. Climate change has been a sustained focus of CMAP's engagement and capacity-building programs since 2017.

Climate Change in Nigeria

Climate change is a global problem, but the effects vary from place to place, based on different climate change threats. According to the IPCC Fifth Assessment Report (AR5), the extent of climate change effects on individual regions will vary over time based on different socio-economic and ecological systems for mitigation and adaptation (2014a). The same assessment stated that Africa is one of the most vulnerable continents to climate change (2014a). This vulnerability of Africa to climate change is driven by several factors of which weak/poor adaptation capacity is one (Ofoegbu and Chirwa 2019). Other studies revealed that the vulnerability of Africa can also be attributed to inadequate access to climate data, institutions, finance resources, and poverty (Ojomo et al. 2015).

Nigeria signed the 2015 Paris Agreement on September 22, 2016 and ratified it on May 6, 2017. Nigeria subsequently developed its nationally determined contributions (NDCs) and submitted it in May 2017. The NDC includes a section on adaptation but focuses primarily on mitigation actions. Nigeria has not submitted a National Adaptation Plan to the UNFCCC; however, it does have a National Adaptation Strategy and Plan for Action which was developed in 2011. This strategy suggests that large urban areas, including Port Harcourt, should develop and implement their own climate change adaptation plans. However, one has not yet been developed for the city.

In order to meet the required implementation of the stipulated provisions under the climate conventions, the department of climate change unit in the Federal Ministry of Environment was created to handle national climate initiatives (Nigerian Conservation Foundation 2008).

In 2003, upon realizing the need to produce observable information on climate change, the Nigerian Meteorological Agency (NIMET) was created, with a department for the development and delivery of climate information.

The majority of analysis on the potential impacts of climate change has been done at the national level. However, in some cases there is evidence of specific impacts in Port Harcourt, Rivers State, and/or the Niger Delta.

Increase in temperature is cited by several sources, with projections that Nigeria may experience as much as 2.5–4.5 °C by 2100 (Akande et al. 2017; Odjugo 2010; FME 2014). Variability in rainfall and the potential for the number of rainy days to decrease throughout the country and around 14% in the Niger-Delta coastal areas has been observed (BNRCC 2011; Odjugo 2005). While the number of rainy days may be decreasing, the frequency and magnitude of wind and rainstorms is increasing (Odjugo 2009).

The rise in sea level and flooding, especially in coastal regions like the Niger Delta, is another major concern (Matemilola 2019; Akande et al. 2017; Duru and Emetumah 2016, 2; Ebele and Emodi 2016, 7; FME 2014, 31; Odemerho 2014). This also impacts agriculture and provision of water due to salt-water intrusion (Odjugo 2010).

Frequent droughts, as the result of climate change, although mostly in the northern part of Nigeria, also have a negative impact on agricultural production and water security (Dioha and Emodi 2018; Elisha et al. 2017; Amanchukwu et al. 2015). Given some of the supply chain connections between Port Harcourt and the north of Nigeria, the impacts may still be felt by residents in Port Harcourt even if the impacts are not projected to take place in the Delta region.

There is some evidence that excessive heat, water stress, and air pollution will have an impact on health in Nigeria, especially in terms of water-related diseases (diarrhea, cholera, and skin diseases), respiratory diseases, exhaustion, skin cancer, and cataract according to one source (Odjugo 2010). Another study found that the four diseases most likely to be exacerbated by climate change in Nigeria are cholera, meningitis, malaria, and pneumonia (Omoruyi and Kunle 2012). A WHO report on climate change and health noted in the country profile on Nigeria that its "peculiarity as one of the leading exporter[s] of oil in Africa, faces [it with] the challenge of balancing global energy demands with the need to address climate and environmental considerations" (WHO 2015).

In Nigeria, the Federal Ministry of Environment defines vulnerability as the element of sensitivity to climate change that relates to how ready a specific system is to respond to changes in the climate, whereas exposure refers to the extent of climate stress to which a particular unit or system is exposed or the presence of people, livelihoods, species, environmental functions, services and resources, infrastructure, or economic, social or cultural assets in places and settings that could be adversely affected (FME 2014).

In an assessment of African cities' capacity to adapt to climate change, Nigeria as a country was ranked 157 out of 192 countries for overall vulnerability based on exposure, sensitivity, ability to adapt, human habitat, and governance readiness (Leal Filho et al. 2018). The ND-GAIN index ranked Nigeria slightly higher, at 127 for vulnerability, but the ranking fell to 148 due to low readiness.

The adaptation gap is the difference between the level of adaptation action in progress and the anticipated needs given the projected climate change impacts. The African Adaptation Initiative's discussion paper from 2018, Enhancing Action on Adaptation in Africa, found that Nigeria is one of the countries with the highest gaps, over 90%.

An analysis by IIED (Satterthwaite et al. 2018) for the 2018 Cities and Climate Change Science Conference assessed likely impacts of climate change specifically on urban populations living in informal settlements and working in the informal economy. This covers a range of climate change impacts that are observed for Nigeria and therefore is relevant to the waterfront communities of Port Harcourt. With higher maximum temperatures, the report notes that the high density, lack of open space, poor ventilation, and corrugated roofs of many informal settlements could contribute to even higher indoor temperatures in these communities. Many informal settlements are in areas at high risk of flooding. These risks are increased by projected intensifications of precipitation. Low-lying waterfront communities in Port Harcourt are clearly particularly vulnerable, and poor quality housing and lack of insurance exacerbate the impact in such communities.

The main outcome of that conference was the Global Research and Action Agenda on Cities and Climate Change Science which highlighted informality as one of six key research gaps as well as the importance of coproducing this knowledge to fill these gaps (World Climate Research Programme 2019). This chapter aims to contribute to the knowledge gap of informality and climate change action as well as the recommendation for the co-production of knowledge as the methodology of the paper, cowritten by community members of the Chicoco Collective with international experts and based on local data collection and global sources reflects this approach.

Climate Change Adaptation: Voices from the Community and Global Recommendations

Global Recommendations for Advancing Adaptation

Adaptation, especially in cities and developing countries, has increasingly become a focus of international organizations and academia, which has resulted in the production of recommendations, guidelines, and priority areas for action.

The recent report Adapt Now: A Global Call for Leadership on Climate Resilience from the Global Commission on Adaptation states that a revolution is needed in three areas to advance adaptation action: understanding, planning, and finance. As part of the revolution on understanding, a better understanding of risks, solutions, and increased knowledge is cited.

In the report's chapter on cities, the main recommendations for the way forward are (1) mainstream information on climate risks in the planning and delivery of urban services, while strengthening local capacity; (2) harness the power of nature to respond to both water and heat risks; (3) build climate resilience by upgrading living conditions in vulnerable communities and informal settlements, drawing on community knowledge; and (4) increase climate-resilient investments and capture value from adaptation benefits (GCA 2019).

An in-depth analysis of several cities around the world and their adaptive capacity determined that cities should focus on the following: the need to prioritize actions, the importance of proper planning, the need for an integrated approach that takes into account structural and development issues, and emphasis on the poorer and more vulnerable communities (Leal Filho et al. 2019).

There has also been an increasing recognition internationally of the need to focus specifically on small and medium-sized cities; however, in a study of European cities, it was found that larger and more prosperous cities are the ones engaged in climate planning, whereas vulnerable cities are less engaged (Reckien et al. 2015). Although the study focused on Europe, the situation in Africa is likely the same and perhaps more so, however data and analysis on this is scarce.

There is not only a discrepancy between types of cities but also on the inequalities within cities and how this affects adaptation. The World Resources Institute (WRI) (2018) highlighted that "[c]ities around the world can thrive when they invest in climate-resilient infrastructure, information management system and risk-reduction programs. But, most often residents who live in risk-prone areas are often left out of the planning and implementation process, leaving them even more vulnerable."

This report highlights that reducing vulnerability is not only about reducing exposure to hazards but also includes a socio-political aspect. Findings from the Urban Community Resilience Assessment program highlight the importance of social cohesion to increase individual and community resilience. This is one reason why relocation can be so damaging to resilience levels. Poor access to information communication systems for early warning and awareness of risks has also been shown to lower resilience (WRI 2018).

Given the importance of public awareness, a report by BNRCC (2011) revealed that the level of public awareness on issues related to climate change in Nigeria is considered to be low. Inadequate information is considered to be one of the key constraints encountered in adapting to climate change in Africa (Otitoju and Enete 2016). Anabaraonye et al. (2019) recommend that information and knowledge about climate change, its causes, and its impacts, as well as mitigation and adaptation strategies, should be made available in simplified, more accessible, forms and translated into local languages. Other technological tools and platforms, such as mobile phone applications and social media, should be utilized in improving information literacy, access to information, and awareness creation (Ayanlade et al. 2017; Duru and Emetumah 2016).

These reports and recommendations suggest some key areas of action that a city like Port Harcourt and organizations like CMAP and the Chicoco Collective could focus on to initiate and strengthen climate change mitigation and adaptation measures, principally: (1) increase understanding about climate change through both the collection and dissemination of locally relevant data and information; (2) carry out long-term and integrated planning; (3) build the local capacity of diverse stakeholders; and (4) prioritize vulnerable communities living in waterfront communities.

Unfortunately, despite the risks and vulnerabilities and the clear need for adaptation measures in Nigeria broadly and Port Harcourt specifically, actions to date have been limited. The reality on the ground presents many challenges to effective action. However, building on the overview of CMAP's work above, the next two sections will further explore the perspectives of local communities and then summarize how the work of CMAP, Chicoco Collective, and local communities can provide a basis for effective action.

Port Harcourt in Focus: Climate Change Adaptation Awareness and Action

The Chicoco Collective devised and conducted a focus group research program to establish a baseline understanding of local perceptions of climate change impacts and adaptation approaches undertaken at the household and community level.

The sessions offered rich insights into local knowledge of climate change, and the various adaptation methods already adopted by informal waterfront communities. The following focus group questionnaire was created to guide the discussion.

Topic	Primary question	Secondary question	Tertiary question	Other question
Climate change awareness	Have you heard about climate change?	What have you heard about		

(continued)

Topic	Primary question	Secondary question	Tertiary question	Other question
		climate change?		
	What do you know about the impacts of climate change?			
Effects of climate change	Have there been changes in your community or city that you can attribute to climate change?	If yes, what are the impacts?	If no, have you experienced extreme heat or cold, flooding, changes in rainfall, more extreme storms, etc.?	
Adaptation	Are you or your household doing anything to adapt to the impacts of climate change? If yes, please explain. If no, why not?	What about your community?	Is the government doing anything?	
Solutions	Can you think of any solutions to adapt to the impacts of climate change?	At the household scale?	Community scale?	Government scale?

The focus group sessions were staged in three different parts of the city. Another session was held for members of the various Chicoco programs. The Chicoco cohort is drawn from waterfront communities across the city. Five men and five women of the Chicoco cohort were randomly selected to participate in that session. Focus group sites were selected to achieve a cross section of neighborhood types and to capture a holistic range of experience across age, socio-economic status, ethnicity, and location. Two communities were chosen from the Old Port Harcourt Township, one of the older colonial areas of the city, which is ringed by over half of, and some of the oldest, waterfront communities in the city. One community was selected from the Diobu axis of Port Harcourt, which is home to over 20 waterfront communities. These communities were selected based on their diverse situations, but also due to existing knowledge of and relationships with community leadership, which facilitated access. Importantly, each of the communities is also host to some aspect of the informal oil and gas industry. The discussions were held in a community space, either a town hall or central area in each community, and participants were recruited

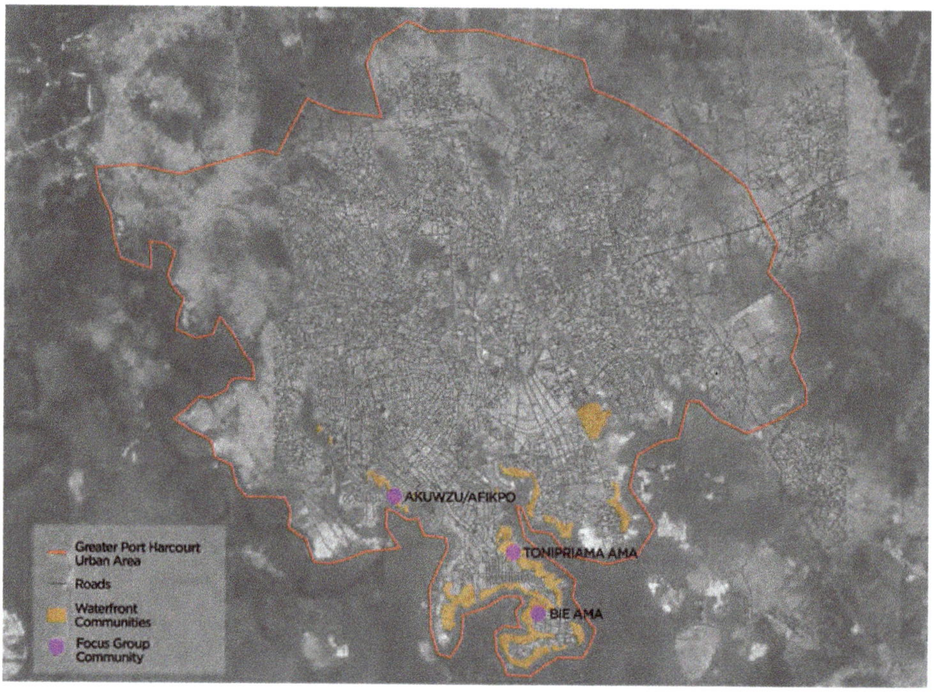

Fig. 10 Map of the greater Port Harcourt highlighting waterfront communities and selected focus group community locations (created by CMAP in March 2020 using QGIS and data from Rivers State Office of the Surveyor General)

with the help of community leadership to include a mix of men and women. A map of Port Harcourt highlighting the focus-group communities can be seen below (Fig. 10).

Akwuzu/Afikpo is a low-income waterfront community in the Diobu area established over 40 years ago. Entrance to the community is through a community market at the top of a steep laterite slope. The community is only accessible by pedestrians on poorly maintained concrete steps. The community has had minimal government-led interventions and minimal community provision of drainage systems, security, schools, and toilet facilities or potable drinking water. Along with other waterfront communities in the Diobu axis, in 2009, Akwuzu/Afikpo was faced with the threat of demolition. Many residents of the community are actively involved in the fishing industry. The community is also a major hub for oil bunkering activities, also known as kpo-fire. The area is predominantly occupied by Kalabari people.

Ten people, five men and five women, attended the focus group session in Akwuzu/Afikpo: the community chief, four fishermen/women, two higher institution students, a retired nurse, and two traders.

Bie Ama Community is a mixed middle- and low-income community in the Old Port Harcourt Township located on a reclaimed wetland area. Bie Ama is made up of five smaller communities or “polos” and is predominantly occupied by Okrikan people. Bie Ama is popularly known as New Road in Borokiri and was established on a sand-fill project in the early 1980s. This community has over 800 structures and has been developed with some regard to a layout plan. It is comprised primarily of cement block structures with aluminum, asbestos, and zinc roofing materials. The community has several churches and a mix of private nursery, primary and secondary schools. Residents include civil servants, teachers, artisans, traders, and contractors. As a waterfront community, the trading of illegally refined or bunkered products is prevalent.

Ten people, seven men and three women, participated in the focus group: two medical doctors, the community youth leader, two bunkerers (workers in the informal oil sector), a retired Nigerian national petroleum company worker, the former Radio Rivers/Nigerian Tide managing director, the female community leader, and two traders.

Tonipriama Ama is the host community of the Chicoco Collective. This waterfront community is comprised of seven smaller communities or "polos." It is predominately occupied by Okrikan people and has existed for over 60 years. This community has over 2,000 structures, constructed from a mix of concrete block and less durable materials such as zinc or wood. The community has a busy commercial dock that is a hub for ferries, fishermen, and oil bunkering boats. There are also several marine companies located in the community. The community has had several government interventions, including the paving of several access roads and the construction of a private government jetty. There are two roads running through the community, and it is easily accessible by motor vehicle. There is one major community public toilet facility and several public boreholes.

The Tonipriama session also comprised ten participants, four women and six men: the Ama (the collective administrative unit containing the polos) chairman, four traders, a teacher, two fishermen, a bunkerer, and a student.

In the focus groups, participants were first asked if they had heard of climate change and what they understood by the term and of the phenomenon. Responses evidenced a general familiarity with the term and an understanding that climate change is related to changing weather conditions and caused by humans. Despite some familiarity with the term climate change, it was evident that many respondents used the phrase to describe any environmental issue, without a clear understanding of the differentiation between degradation (air quality issues, water pollution, etc.) and changes in climate (changes in precipitation, flooding, increased temperature, etc.).

Climate Change Impacts Observed

All of the communities shared experiences that they attributed to climate change and the sense that its impacts are felt citywide, increasing in intensity and frequency. Heat and flooding (both from sea level rise and excessive rainfall) were raised across groups as an observable impact of climate change, popular perceptions aligning with current scientific research on impacts of climate change in Nigeria.

Food insecurity and hunger, as a result of food scarcity, increased food prices, difficulty of farmers to predict the weather for planting seasons, scarcity of seafood due to decrease in catch or the catching of dead fish, and food storage and spoilage issues were raised by all groups. More research is needed to ascertain what links, if any, these issues have to climate change. However, these are clearly primary concerns for communities and should be addressed in adaptation and/or development research and action.

Health issues were also raised by all groups: Disease burden increase, with measles and chicken pox cited specifically, rashes, dehydration, respiratory issues, cancer, and eye conditions were all perceived by participants as being linked to climate change. While such attribution is far from clear for all the conditions mentioned, connections to air pollution seem to be a reasonable assumption, and it is likely that the dehydration and rash conditions are related to heat and/or water quality and access. More research would be needed to confirm such links. Nevertheless, it is clear that here too, adaptation and development measures would need to address community-identified health priorities.

The issues of pollution and environmental degradation were another focus of all groups. The black soot crisis has gained citywide attention recently, and this was reflected in the group discussions. The toxicity of rainwater, rendering it unusable for consumption or cooking, was mentioned by many. Marine habitat pollution and its impact on fish stocks were another commonly voiced concern for waterfront communities with members who still depend directly on aquatic urban ecosystems for their livelihoods. Again, while air and water quality may be linked to climate change-driven dynamics, not all of these issues are necessarily directly attributable to climate change. Nevertheless, it is likely that climate change exacerbates these problems. It is clear that environmental degradation is a major concern and in Port Harcourt the drivers of climate change are the same causes of environmental degradation. Therefore both mitigation and adaptation efforts should address these environmental issues.

Other issues raised in different groups, relating to migration, cost of living increases, and security, may involve interesting direct or indirect relationships to climate change. For instance, migration is a much researched topic in this connection, but more area-specific data would need to be gathered to determine causality and appropriate adaptation actions.

Adaptation Actions

As the effects of the climate crisis intensify, residents of Port Harcourt have developed different coping techniques for them. The participants of the focus group highlighted some of the ways they or their neighbors are changing their behavior. For increased heat and rising temperatures, coping techniques included: bathing with cold water three to four times a day, showering multiple times at night, and sleeping outside till late in the night to get some cooler air before going indoors and then sleeping on relatively cool mud or stone floors.

Increases in water intake to mitigate dehydration caused by the heat have significant financial impacts on households that spend disproportionately high proportions of their income on food and water essentials. Participants reported that water is also used to keep food cool and to slow its spoiling in the absence of reliable power for refrigeration. Other practices of drying food with salt were also emphasized as a way to preserve their food and kill pests, such as worms. Additionally, as a result of increased food prices, some respondents say that they are eating only once a day. Self-rationing is increasingly common.

Adaptation methods to address flooding include improving waste management by methods such as establishing more designated dumping sites to discourage people from throwing waste in waterways. Other discussed methods included community gutter cleaning efforts, construction of high tables to store personal belongings, blocks to raise beds, and raised doorsteps to prevent water ingress.

Respondents discussed their discomfort with and concern for the black soot air quality crisis facing their city. Concern for their health has prompted some to start wearing facemasks on days when air quality was visibly bad. The wearing of facemasks was practiced before the recent coronavirus pandemic.

Despite being asked about their adaptive practices, several of the responses were not necessarily relevant as adaptation measures, but really more about other impacts related to a changing climate, particularly increasing temperatures. Many focus group participants brought up the issue of increased lethargy and unproductivity at work from the higher temperatures. Others mentioned the need to increase their use of generators, which although not explicitly said, is likely related to needing to run fans.

It is clear that most adaptation efforts are at an individual level or household level. Knowledge of adaptive practices at community or broader scale was quite limited. But, as noted in the findings from the fecal sludge management study, some communities are undertaking collective adaptation actions, such as managing flooding issues in a coordinated manner with berms and nets. It is unclear as to whether focus group participants are simply unaware of any actions that their community or the government may be undertaking or whether they are not able to relate any actions specifically to a response to a changing climate. This highlights the need for more awareness raising of potential adaptation actions as well as planning for more coordinated efforts.

One of the communities is making efforts to curb issues of poor sanitation and dumping by discouraging the act of disposing waste in waterways by fining offenders. Two of the communities have active community leadership committees and subcommittees in charge of sanitation, and community electricity and light, among other specialized issues. These committees are tasked with maintaining community infrastructure and solving challenges that arise in the community by enacting rules and regulations, sometimes including fines or punishments for breaking these rules, or introducing levies to construct new infrastructure. Tonipriama Ama, for example, has been charging for use of and maintaining a public toilet, and likewise through its light committee, has organized levies to replace light poles and broken transformers. Although currently these committees have not addressed impacts directly related to a changing climate, these initiatives demonstrate a capacity to develop and deploy adaptations at community level.

Proposed Adaptation Solutions

When asked about other potential solutions for addressing climate change, the participants highlighted the need for education, behavior change, infrastructure projects, and government intervention. They proposed that, to tackle the effect of the climate crisis, government and relevant stakeholders needed to be more responsive, accountable, and active. They criticized lack of action, insufficient resourcing, and lack of human and technical capacity. Government and relevant stakeholders were urged to lead sensitization efforts and to provide basic amenities. There was a clear understanding of interplay between behavior and infrastructure. For example, on the one hand, the need for changed waste dumping practices, particularly regarding waterways, and for waste disposal practices that facilitate recycling and reuse was recognized. On the other hand, groups also stressed the need for better waste management systems including the placement of proper disposal containers within the communities and the construction of better gutters and drainage systems.

Groups proposed awareness-through-action and learning-by-doing approaches, as well as the need for more active community debate. Tree planting and community town hall programs were mentioned in this connection. Community sanitation exercises – the mobilization of the entire community to clean streets, drains, and public spaces – were also mentioned. Indeed, this capacity to mobilize, *en masse, as* a community is one of the significant resources that such settlements can draw on.

Energy was a perennial topic of debate. Gas flaring has long been a highly visible example of the violent environmental externalizations of the hydrocarbon economy in the Niger Delta. People linked the demand to end flaring with the desire to transition to renewable energy. Respondents wanted to see oil and gas companies halt polluting activities and proposed "trapping" carbon emissions. The technical approaches to doing so were not detailed. Security forces tasked with policing the informal oil industry often spill confiscated hydrocarbon products directly into waterways or set them ablaze. Respondents voiced strong opposition to these practices and proposed that, instead, confiscated product should be taken to official refineries. The retraining of bunkerers and the regulation of the informal economy to improve the environmental standards of operations were also proposed. While these are not strictly adaptation efforts, they can be considered as modest mitigation efforts and community actions aimed at a cleaner community environment.

Conclusions and Recommendations

Port Harcourt and many other cities in Africa face a multitude of climate change, environment and development challenges. These challenges are greatest in informal settlements. As informality is the dominant mode of African urbanism, it is critical that interventions respond to the situated constraints and potentials of informality and how it is variously manifested in communities and neighborhoods across the continent. To date, climate change adaptation has not necessarily been a priority in and for informal urban settlements. Many of the requirements for effective action –

from data to finance – are often lacking. Too often, the priorities, knowledge, skills, and participation of local residents themselves are missing from external interventions. However there is growing awareness of the problems and initial steps and information that can be built upon for effective solutions. Below are recommendations drawn from the experience of CMAP, Chicoco, and the waterfront communities in Port Harcourt. As the formulation of these recommendations has been led by residents and through partnership with international experts, we hope they might serve as a basis for mobilization and action in other African cities facing similar constraints and opportunities.

- *Awareness and local knowledge.* There is a of lot consistency between scientific findings on the impacts of climate change in Nigeria and those observed and recorded in local focus groups. This suggests that, while innovation is needed in the popular dissemination of research findings, many residents in informal settlements may not be aware of the latest research on climate change, they are acutely aware of direct impacts from their own lived experience and are developing local adaptations. This experience, critical awareness, and approach to situated adaptation are a potentially valuable asset for climate change response efforts.
- *Building awareness through local methods.* International organizations and governments should draw on locally established information sharing and storytelling initiatives and networks. Chicoco Radio and Chicoco Maps' Encounter and Exchange program are examples of the potential of community media and action programs not only to raise awareness and strengthen networks locally, but also to catalyze citywide dialogue. The roadshows, concerts, town hall listening clubs, broadcasts, and live discussion programs that were part of Chicoco's first radio drama season successfully stimulated citywide debate around issues of forced evictions and police brutality. Such a platform could easily be mobilized for climate change adaptation and environmental justice issues.
- *Inclusive data and knowledge collection and sharing.* There is a need for more institutional investment in knowledge and data gathering efforts on climate change impacts in Nigeria at the national and local levels. To collect and disseminate knowledge and data relevant to informal urban settlements in Nigeria, residents of those settlements need to be successfully involved. The deep, detailed, accurate, and appropriate datasets gathered by the young Chicoco Maps team with CMAP's support demonstrate a successful methodological approach to, and effective methods of, participatory data gathering and sharing in Nigeria's informal settlements. This participatory, partnership-based approach can be mobilized in a more focused way on climate change research and the development of sustainable local adaptations.
- *Integrated solutions and local context.* Adaptation solutions need to tackle development and environmental justice issues alongside climate change adaptation action, especially health issues. It is also important to recognize the context of the local economy – both formal and informal. In the case of Port Harcourt, this includes reckoning with the economic domination of the oil and gas industry.

- *Small and local efforts can be compounded for larger impact.* Some routes to scale involve a concert of small-scale interventions. CMAP and Chicoco Maps' citywide sanitation research found that the solution to Port Harcourt's sanitation crisis is not a billion-dollar central sewerage system. Rather, onsite community solutions, manageable at neighborhood-level and network-scalable, offer a sustainable approach through which many small actors can coordinate to achieve scale. As found in the focus groups, individuals and households will take adaptive measures that can be effective for reducing loss and vulnerability.
- *Large-scale responses are also necessary, with community involvement and input.* Local communities have devised some effective individual or neighborhood-level adaptation techniques, but larger infrastructure interventions, requiring access to significant capital and technologies, are necessary. Yet, for these solutions to achieve sustainability, they nonetheless need to engage and involve community actors. In a city like Port Harcourt, where the government suffers a deep trust deficit, large-scale interventions often fail because of a lack of local cooperation or even active resistance. The trust and involvement of community actors is crucial to large-scale climate change responses, particularly as large-scale behavior changes and cultural shifts are necessary for success.
- *Bottom-up collaboration to build accountability.* A bottom-up approach follows from the previous recommendation. While community members have raised the need for government intervention and effective government service provision, heavily top-down approaches to external interventions in communities with already strained relationships with their representatives are going to be difficult. Involving communities, not merely in consultation exercises, but actively in project design, is key. This is likely both to increase the appropriateness of adaptation measures, and importantly, to build a robust demand for duty bearer accountability. Demand-side platforms for accountability for climate change action in cities like Port Harcourt are critical to counter a culture of flourishing impunity that has allowed staggering environmental degradation and unsustainable exploitation to go unchecked for decades.

Acknowledgments In-kind financial support was provided by the Chicoco mapping team for research, data collection and analysis, and the focus group discussion support team, namely the recording technician, George Sosipiriala, and community engagement specialist, Fubara Tokuibiye Samuel. Michael Uwemedimo, Ana Bonaldo, and the rest of the CMAP team also provided in-kind support and guidance for this chapter.

References

African Adaptation Initiative (2018) Enhancing action on adaptation in Africa. Discussion paper for Africa state of adaptation report

Akande A, Costa AC, Mateu J, Henriques R (2017) Geospatial analysis of extreme weather events in Nigeria (1985–2015) using self-organizing maps. Adv Meteorol 2017(2):1–11

Amanchukwu RN et al (2015) Climate change education in Nigeria: the role of curriculum review. Education 5(3):71–79

Anabaraonye B et al (2019) Educating farmers and fishermen in rural areas in Nigeria on climate change mitigation and adaptation for global sustainability. Int J Sci Eng Res 10(4):1391–1398

Ayanlade A et al (2017) Comparing smallholder farmers' perception of climate change with meteorological data: a case study from southwestern Nigeria. Weather Climate Extremes 15:24–33

Bloch R, Monroy J, Fox S, Ojo A (2015) Urbanisation and urban expansion in Nigeria. Project report. ICF International, London

BNRCC (Building Nigeria's Response to Climate Change) (2011) National adaptation strategy and plan of action on climate change for Nigeria (NASPA-CCN). Prepared for the Federal Ministry of Environment Special Climate Change Unit

Collaborative Media Advocacy Platform (CMAP) (2018) Ibiapu Polo, Amatari Polo, and Darick Polo. Household Survey Data Summary Report

Dioha MO, Emodi NV (2018) Energy-climate dilemma in Nigeria: options for the future. IAEE Energy Forum 2018(2):29–32

Duru NP, Emetumah CF (2016) Evaluating the effects of information literacy on climate change awareness among students in Imo State University. Arch Current Res Int 4(3):1–10

Ebele NE, Emodi NV (2016) Climate change and its impact in Nigerian economy. J Sci Res Rep 10 (6):1–13

Elisha I et al (2017) Evidence of climate change and adaptation strategies among grain farmers in Sokoto State, Nigeria. IOSR J Environ Sci Toxicol Food Technol (IOSR-JESTFT) 11(3):1–7

Federal Ministry of Environment (FME) (2014) United Nations climate change Nigeria. National Communication (NC). NC 2. 2014

Global Commission on Adaptation (2019) Adapt now: a global call for leadership on climate resilience, Washington, DC

IPCC (2014a) Summary for policymakers. In: Field CB et al (eds) Climate change 2014: impacts, adaptation, and vulnerability. Part A: global and sectoral aspects. Contribution of working group II to the fifth assessment report of the intergovernmental panel on climate change field. Cambridge University Press, Cambridge/New York, pp 1–32

Leal Filho W et al (2018) Strengthening climate change adaptation capacity in Africa-case studies from six major African cities and policy implications. Environ Sci Pol 86:29–37

Leal Filho W, Balogun AL, Olayide OE et al (2019) Assessing the impacts of climate change in cities and their adaptive capacity: towards transformative approaches to climate change adaptation and poverty reduction in urban areas in a set of developing countries. Sci Total Environ 692:1175–1190

Matemilola S (2019) Mainstreaming climate change into the EIA process in Nigeria: perspectives from projects in the Niger Delta region. Climate 7(2):29

Nigerian Conservation Foundation (2008) Climate change. http://www.ncfnigeria.org/component/content/article/90-programmes/127-Climate-Change. Accessed Mar 2020

Odemerho FO (2014) Building climate change resilience through bottom-up adaptation to flood risk in Warri, Nigeria. Environ Urban 27(1):139–160

Odjugo PA (2005) An analysis of rainfall patterns in Nigeria. Glob J Environ Sci 4(2):139–145

Odjugo PAAO (2009) Quantifying the cost of climate change impact in Nigeria: emphasis on wind and rainstorms. J Hum Ecol 28(2):93–101

Odjugo O (2010) General overview of climate change impacts in Nigeria. J Hum Ecol 29(1):47–55

OECD/SWAC (2020) Africapolis (database), www.africapolis.org. Accessed 11 Feb 2020

Ofoegbu C, Chirwa PW (2019) Exploring the potential for green growth uptake in the South African forest sector. Reg Environ Chang 19(5):1469–1480

Ojomo E, Elliot M, Amjad U, Bartram J (2015) Climate change preparedness: a knowledge and attitudes study in southern Nigeria. Environment 2(4):435–448

Omoruyi EP, Kunle OA (2012) Effects of climate change on health risks in Nigeria. Asian J Bus Manag Sci 1(1):204–215

Otitoju MA, Enete AA (2016) Climate change adaptation: uncovering constraints to the use of adaptation strategies among food crop farmers in South-west, Nigeria using principal component analysis (PCA). Cogent Food Agric 2(1):1–11

Reckien D, Flacke J, Olazabal M, Heidrich O (2015) The influence of drivers and barriers on urban adaptation and mitigation plans—an empirical analysis of European cities. PLoS One 10(8):1–21

Satterthwaite D, Archer D, Colenbrander S, Dodman D, Hardoy J, Patel S (2018) Responding to climate change in cities and in their informal settlements and economies. Consultant report presented at the IPCC international scientific conference on cities and climate change, Edmonton. International Institute for Environment and Development, London

Theis M, Lloyd-Jones T, Adenekan S, Gusah S, Moor M, Mulyawan B, Gusah D (2009) Port Harcourt waterfront urban regeneration: scoping study. Project Report. MLC Press, Max Lock Centre, University of Westminster, London

World Bank (2016) From oil to cities: Nigeria's next transformation. Directions in development. World Bank Group, Washington, DC

World Bank (2017a) Sustainable WSS services in Nigeria (Vol. 2): Data collection report (English). World Bank Group, Washington, DC

World Bank (2017b) Sustainable WSS services in Nigeria: fecal sludge management – a practical guide for evaluating needs and developing solutions (English). World Bank Group, Washington, DC

World Climate Research Programme (2019) Global research and action agenda on cities and climate change science – full version. In: Prieur-Richard AH, Walsh B, Craig M, Melamed ML, Colbert M, Pathak M, Connors S, Bai X, Barau A, Bulkeley H, Cleugh H, Cohen M, Colenbrander S, Dodman D, Dhakal S, Dawson R, Espey J, Greenwalt J, Kurian P, Lee B, Leonardsen L, Masson-Delmotte V, Munshi D, Okem A, Delgado Ramos GC, Sanchez Rodriguez R, Roberts D, Rosenzweig C, Schultz S, Seto K, Solecki W, van Staden M, Ürge-Vorsatz D (eds) WCRP Report No. 3/2019, 31pp

World Health Organization (2015) Climate change and health: country profile. United Nations Framework Convention on Climate Change, Nigeria

World Resources Institute (2018) Prepared communities: implementing the Urban Community resilience assessment in vulnerable neighborhoods in three communities, Washington, DC

5

Role of Cassava and Sweetpotato in Mitigating Drought in Semi-Arid Makueni County in Kenya

C. M. Githunguri and E. N. Njiru

Contents

Abstract

Cassava and sweetpotato are major factors in food security across sub-Saharan Africa. Though cassava and sweetpotato varieties that are early maturing and resistant to diseases have been developed, many farmers still grow local varieties. Cassava and sweetpotato cultivars that mature between 6 and 12 and 3 and 4 months after planting, respectively, are available. The objective of the synthesis was to obtain a general overview of cassava and sweetpotato production in Matiliku subcounty of Makueni County in semi-arid eastern Kenya before the establishment of a seed system for them. Participatory rural appraisal and focused group discussions with key stakeholders in Makueni County on the current status

C. M. Githunguri (✉)
Kenya Agricultural and Livestock Research Organization (KALRO) Food Crops Research Centre Kabete, Nairobi, Kenya

E. N. Njiru
KALRO Katumani, Machakos, Kenya
e-mail: elias.njiru@kalro.org

of these crops provided very useful information. It was observed that there are a few early cassava and sweetpotato adopters, meaning a lot of effort in communicating the need to commercialize them needs to be made. Even though the farmers had sufficient experience in growing them at subsistence level, they were searching for cultivars that combine both nutritional and food security. There is a need to engage more extension service providers in order to campaign on their adoption. There is a need to carryout training and awareness creation on their role in food security and wealth creation.

Keywords

Cassava · Food security · Kenya · Makueni · Sweetpotato

Introduction

Cassava and sweetpotato produce starchy tuberous roots. Cassava and sweetpotato may be sweet or bitter depending on the variety. The crops are resistant to drought and are mainly propagated through stem cuttings or vines. Cassava and sweetpotato can adapt to diverse climatic conditions, survive long dry spells, and can be harvested and stored on a flexible time schedule, all of which qualifies cassava and sweetpotato as food security crops and technologies that respond well to climate change. Cassava and sweetpotato tuberous roots are rich in carbohydrates and are a staple food for many Africans. They are also used to manufacture alcohol. The leaves are richer in protein and minerals than the tuberous roots both qualitatively and quantitatively and are used both as vegetable and fodder. Cassava and sweetpotato have high potential if people are sensitized to their usefulness and nutritive value. Currently, women and children usually grow cassava and sweetpotato as food security crops. However, the scenario changes when the two crops are grown for commercial purposes with men playing a prominent role (Nweke et al. 2002).

Thirty-centimeter (30) stem cassava cuttings are planted upright, slanting, or buried horizontally. Sweetpotato is propagated also from 30 cm vines, about 2/3 s of which are inserted in the ground. Both crops have a great yield potential of 20–50 t/ha of root dry weight in the tropics. This yield potential has yet to be exploited as farmers are producing an average of less than 10 t/ha. Cassava and sweetpotato provide 8% or more of the minimum calories requirements of some 750 million people in the tropics, making it one of the most important energy sources in the human diet. Cassava and sweetpotato have great potential as industrial starch production crops and hence provide employment at the production, marketing, and sales levels. Given that the demand for food and feed remains unsatisfied and the time pressure to meet the challenge is increasing rapidly, cassava and sweetpotato can help alleviate this problem.

Cassava and sweetpotato produce about 10 times more carbohydrates than most cereals per unit area and are ideal for production in marginal and drought prone areas, which comprise about 75% of Kenya (Githunguri et al. 1998; Githunguri

2002; Nweke et al. 2002). Despite their great potential as a food security and income generation crops among rural poor in marginal lands, their utilization remain low. The potential to increase cassava and sweetpotato utilization is enormous with increased recipe range (Githunguri 1995). The International Institute of Tropical Agriculture (IITA) has officially recognized cassava as a new cash commodity and a vital food in Africa. The Common Fund for Commodities has recognized cassava as an internationally tradable commodity. The Intergovernmental Group on Grains has adopted cassava as a commodity. Cassava and sweetpotato can accelerate development and improve livelihoods of Kenyan rural communities as has been demonstrated in Vietnam and Thailand. Adoption of improved cassava production, processing, and marketing technologies earns these countries about one million US$ per annum. This model is being pursued by initiatives in Africa such as the NEPAD Pan African Cassava Initiative, Sub-Saharan Africa Challenge Program, and the South African Root Crops Research Network and East African Root Crops Research Network.

Cassava and sweetpotato are major factors in food security across sub-Saharan Africa. In Kenya, cassava is grown in over 90,000 ha with an annual production of about 540,000 tons. Cassava (*Mahihot esculenta* Crantz.) is a relatively neglected tropical root crop in East and southern Africa (Githunguri 1995; Githunguri et al. 2017a, b, c). It is grown widely in East Africa in areas below 1500 m above sea level (Acland 1985). Cassava production in Kenya is concentrated in three main regions: Coastal, Central, and Western region. Western and Coastal regions are the main cassava producing areas, producing over 80% of the recorded cassava output in the country (Githunguri et al. 2017a). Though cassava is considered to be a food security crop in the sub-Saharan Africa, its production in Kenya is low compared to other crops like maize, beans, and sorghum. Its consumption is low especially in the central region of Kenya where it is considered a poor man's crop and is usually consumed during periods of food scarcity. Despite its high production in the coastal and western regions of Kenya, utilization is limited to human consumption. In order to promote production, which has been decreasing in recent years, there is a need to explore and identify other uses of cassava.

According to FAO (1990), Africa produces about 42% of the total tropical world production of cassava and contributes significantly to food security across sub-Saharan Africa. Githunguri et al. (1998) and Nweke et al. (2002) noted that cassava could grow well in marginal lands, requires low inputs, and is tolerant to pests and drought. Cassava is grown in over 90,000 ha with an annual production of about 540,000 t in Kenya and could remain in the ground for 7–24 months after planting and then harvested thereafter (Githunguri et al. 2017a). Utilization of cassava remains low in Kenya because the fresh root tubers are limited to roasting and boiling for consumption only despite its great potential as a food security and income generating crop (Githunguri 1995; Githunguri et al. 2017a). However, it should be noted that cassava is widely used in Kenya by almost all communities despite the fact that there is still a lot of room for expansion on its use especially by industrialists who have yet to fully utilize cassava in food and stock feed manufacture. The Home Economics Department of the Ministry of Agriculture and other organizations have a wide range of options in developing cassava recipes acceptable to a larger community (Githunguri 1995; Githunguri et al. 2017a). In Kenya, cassava is the second most important food root crop after *Solanum* potato. However, due to its small production base, it is ranked 36th out of 50 in KALROs 1991 priority setting exercise (KALRO 1995; Githunguri 1995; Githunguri et al. 2017a). There is a slow but steady increase in cassava production because its consumption has a direct effect on demand.

Variability studies by Rai et al. (1986) on cassava showed that phenotypic variance was higher than genotypic variance for all characteristics, which included plant height, tuberous root girth, length, and weight. Later rain-fed trials involving several cassava cultivars showed significant genotype by environment (g × e) interaction for yield (Naskar et al. 1989). Selection for a slightly higher than optimal leaf area index, and hence greater biomass, can lead to stable cassava yields across both favorable and stress environments. Further, according to Nartey (1981) and Githunguri et al. (2014), the harvested cassava yields depend on the distribution of dry matter between the useful and other parts of the plant, which may be affected by the sink capacity of the useful parts to accept photosynthates, as well as the capacity of the leaves to supply it. In sub-Saharan Africa, most boil and eat fresh cassava consumers prefer roots with high dry matter contents, which are boiled and consumed as vegetable (Nartey 1981; Githunguri et al. 2014). It is important to note that Dixon et al. (1994) had reported negative correlation between root dry matter content and root cyanogenic potential which is important when selecting cassava varieties targeting fresh tuberous root consumers. According to IITA (1990b) and Githunguri et al. (2016), useful yields of the cassava crop depend on its total biomass and its distribution among useful parts. As such, since the total light intercepted strongly depends on leaf area, the best yielding cassava is one that establishes adequate and efficient photosynthetic surface sufficiently early in the cropping season, maintains it at least over the duration of the moisture cycle, and partitions dry matter into yield (Cock 1985); IITA 1990a). Other studies (IITA 1990b) have shown that best root yields are obtained at a leaf area index of 3.5, with general reduction in yield as the leaf area indices increase.

African farmers usually grow cassava under field conditions where one or more of the resources are limiting. However, it should be noted that most of the research works carried out are under optimum management conditions (Githunguri et al. 2006). Importance of cassava as an industrial crop is increasing rapidly in several countries within the tropics and especially sub-Saharan Africa (Ayinde et al. 2004; Azogu et al. 2004; EFDI-Technoserve 2005; Ezedinma et al. 2005; Githunguri 1995; Onyango et al. 2006). According to several studies (Githunguri 2002; Odongo 2008; Tivana and Bvochora 2005), safety of cassava products is important and could be affected by agro-ecological zones and genotypes. According to Ekanayake et al. (1997a) and Ekanayake et al. (1997b), a cassava plant possesses several physiological parameters and processes that confer to it the ability to produce modest yield under a range of both abiotic and biotic stresses. Some of these parameters include heliotropism, leaf drooping, long fibrous roots, loss of leaves, high water use efficiency, and large source potential, and sink capacity (Ekanayake 1998; Githunguri 2002). Other studies have shown that the cassava value chain has three main components: field production, processing, and marketing (Githunguri et al. 2017a, b, c). Though cassava varieties that are early maturing and resistant to diseases have been developed, many farmers in the cassava growing regions (coastal, central, and western) still grow local varieties (Githunguri et al. 1998, 2006, 2007a; Githunguri 2002). In addition, early maturing, high yielding, disease, and pest tolerant cassava cultivars developed through participatory breeding by the Kenya Agricultural and Livestock Research Organization (KALRO) are available to farmers (Githunguri et al. 2006, 2007a).

Sweetpotato is an important staple food in the East African region after banana and maize. It has high potential for livestock feed and industrial use as well (Onwueme 1978). Generally, small-scale farmers using traditional farming systems on marginal soils produce most of the sweetpotatoes and cassava. According to Onwueme (1978), yields of sweetpotato and cassava vary with cultivar, disease resistance, and location and production practices. Constraints to sweetpotato production in Kenya include the sweetpotato weevil, lack of adequate disease and pest-free planting materials, poor cultural practices, lack of appropriate storage and processing technologies, and poor market infrastructure (Lusweti et al. 1997; Githunguri et al. 2003). Although some trade occurs, there is minimal official record of international trade in sweetpotato. Kenya mainly meets her consumption needs almost exclusively through internal production. Thus, sweetpotatoes like cassava are consumed locally and hence traded in a closed economy. This suggests that prices are mainly dependent on local forces of supply and demand. Due to lack of external market, the aggregate demand for sweetpotato within Kenya is assumed to equal supply (Odendo et al. 2001). One of the major constraints towards cassava and sweetpotato production and commercialization is their low and slow multiplication ratio due to their being vegetatively propagated. This constraint is further compounded when their production is carried out in the arid and semi-arid areas where their multiplication can only be effectively and commercially multiplied under irrigation and high inputs. Most local cassava and sweetpotato cultivars are low yielding and are late maturing. Though cassava and sweetpotato varieties that are early maturing and resistant to diseases have been developed, many farmers still grow local varieties. KALRO has sweetpotato cultivars that mature between 3 and 4 months and cassava cultivars that produce appreciable yields between 6 and 12 months after planting. As such, the specific objective of this synopsis was to obtain a general overview of cassava and sweetpotato production in Matiliku Division Makueni District in semi-arid eastern Kenya before the establishment of a seed system for the two crops.

Farmer Interviews

Participatory rural appraisal (PRA) and Focused Group Discussions with farmers, local leaders, extension staff, and other stakeholders were conducted at Matiliku Division in Makueni District. The PRA involving 51 farmers was conducted at a farm with an elevation of 1043 m above sea level located on Latitude S: 011.867812 and Longitude E: 042.78234. Using a semi-structured questionnaire, farmers were interviewed individually whereby an overview on cassava and sweetpotato production in Matiliku was obtained. The data was subjected to descriptive analysis methods while charts were generated using Microsoft Excel application. The results were used to inform decisions on their training and cassava and sweetpotato production requirements.

Survey Observations and Synthesis in a Semi-Arid Area in Kenya

Figure 1 shows about 35.3% of the farmers put 0.25 acres under cassava cultivation while 13.7% have a few stands and 33.3% do not grow cassava at all. About 11.8% of the respondents put in cassava on land ranging from 0.5 to 1.5 acres and this category could be the early adopters.

The number of years under cassava production ranged between 1 and 40 years with only a few (7.7%) farmers indicating to have been growing cassava for a period above 40 years (Fig. 2). The majority of farmers, about 41.1%, have grown cassava

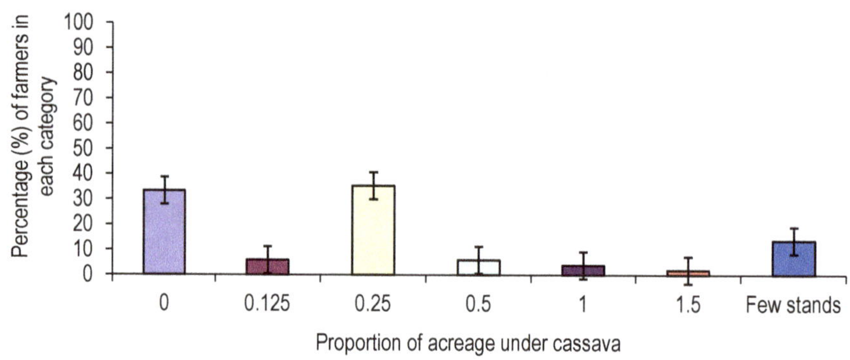

Fig. 1 Proportion of acreage under cassava production in Matiliku Division, Nzaui District, Makueni County. Vertical bars represent the standard error between means ($P = 0.05$)

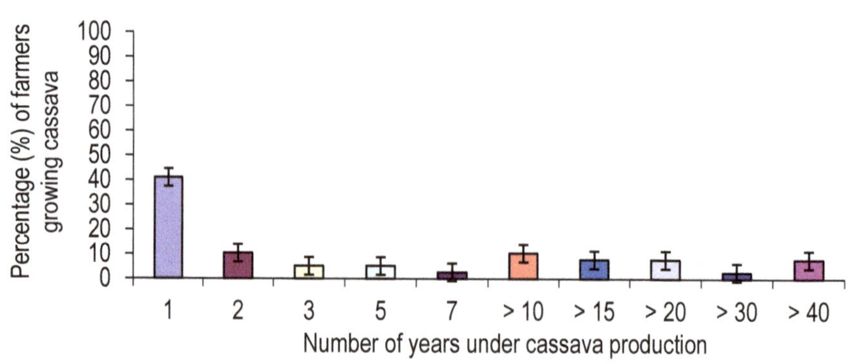

Fig. 2 Number of years under cassava production and proportion of farmers growing cassava in Matiliku Division, Makueni (Nzaui) District. Vertical bars represent the standard error between means ($P = 0.05$)

for only 1 year while the rest had cassava-growing experience ranging between 2 and 30 years in proportions ranging between 5.1 and 10.3%. These results suggest that cassava is grown as a subsistence crop and a lot needs to be done if it is going to be commercialized in Matiliku.

According to Table 1 the majority of farmers, about 40.5%, were growing a local cassava cultivar whose name seemed to be unknown while 31.0% were growing a cultivar whose origin was uncertain. Among the known local cultivars, Kitwa was being grown by 7.1% of the respondents. A similar percentage (7.1%) of respondents were growing the improved "Mucericeri" cultivar obtained from KALRO and FIPs Africa. Other minor cultivars included 990,005, "Kiisungu," and "Nyawo." These results suggest that even though KALRO and some significant service providers have started distributing improved cultivars, a lot still needs to be done if cassava is going to play its rightful role in supplying the much needed calories.

Table 2 shows that the most significant methods of utilization include boiling and eating cassava for breakfast and chewing raw as a snack. About 10.1% of respondents mix cassava with maize and beans ("githeri") while 17.7% grow cassava for

Table 1 Types of cassava cultivars cited and proportion of farmers growing them in Matiliku Division, Makueni (Nzaui) District

Types of cassava cultivars cited	Percentage (%) of farmers growing them
990,005 (improved)	2.4
FIPS Africa (improved)	2.4
(KALRO) improved	4.8
Kiisungu (local)	2.4
Kitwa (local)	7.1
Local (unknown)	40.5
Mucericeri (improved)	7.1
Nyawo (local)	2.4
Unknown (not certain improved or local)	31.0

Table 2 Method of utilization of cassava and proportion of farmers involved in Matiliku Division, Makueni (Nzaui) District

Method of utilization	Percentage (%) of farmers utilizing them
Boil and eat as a snack	2.5
Boil for breakfast	35.4
Chew raw	24.1
Give to friends	1.3
Have not utilized	5.1
Mix with "Githeri"	10.1
Mix with meat to make a stew	1.3
Peel, chip and mix with sorghum or maize	1.3
Process cakes	1.3
Sell in the market	17.7

sale in the local market. Only 5.1% of the respondents indicated they have not utilized cassava in any way. The results indicate that there is very little processing of cassava taking place in Matiliku suggesting that there is need to carryout training and awareness creation on the role of cassava in food security and wealth creation.

Results in Table 3 show that KALRO (35.9%) has been largely responsible in the introduction of cassava as a viable technology in this area. 25.6% of the respondents were not certain about the origin of their cultivars, whereas 20.5% obtained them from the local market. It seems 20.5% of the respondents purchased cassava cuttings from the market, which is a positive sign on the viability of cassava being taken up as a commercial venture. There is need to increase the number of extension service providers in the area.

Figure 3 shows that the majority of respondents comprising 52.5%, 30%, and 10% grow sweetpotatoes on 0.25, 0.125 acres of land and only a few stands, respectively. The rest 7.5% grow sweetpotatoes on pieces of land ranging from 0.5 to 2 acres who seem to comprise the group of early adopters. These results suggest that sweetpotato is also grown as a subsistence crop. Serious efforts need to be put in place in order to move sweetpotato from subsistence level to commercialization in Matiliku.

Table 3 Origin of cassava cultivars cited and proportion of farmers growing them in Matiliku Division, Makueni (Nzaui) District

Origin of cassava cultivars cited	Percentage (%) of farmers growing them
Farmers	7.7
FIPs Africa	7.7
Inherited from parents	2.6
KALRO	35.9
Local market	20.5
Unknown	25.6

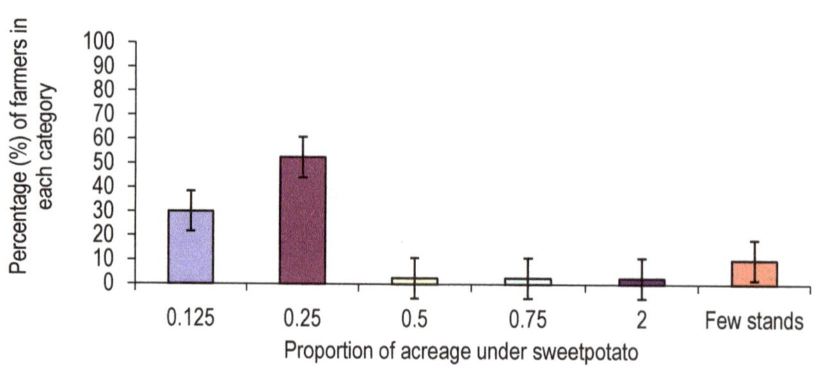

Fig. 3 Proportion of acreage under sweetpotato production in Matiliku Division, Makueni (Nzaui) District

Figure 4 shows that majority of the respondents, about 60%, have mainly grown sweetpotatoes for the last 3 years. The rest 40% have mainly grown sweetpotato for a period ranging from 4 to 30 years. This suggests that the concerted efforts that have been put in place by KALRO and its partners in extension service providers during the last 4 years have paid divided and all that is needed is to sustain the efforts.

The results in Table 4 show that 21.2% of the respondents were growing a local cultivar of an unknown origin while 16.7% of the farmers were growing an unknown improved cultivar presumably obtained from KALRO or other extension service providers. Sallyboro, an improved orange-fleshed sweetpotato cultivar, and "Mwezi moja," a drought tolerant local cultivar, were the only cultivars being grown by an appreciable number of respondents, 15.2% and 10.6%, respectively. The rest of the

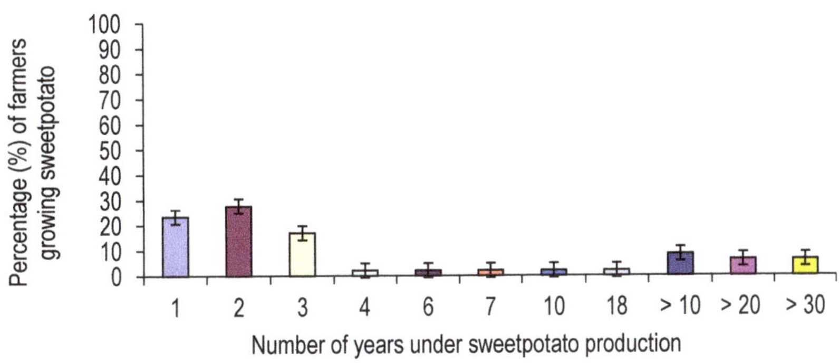

Fig. 4 Number of years under sweetpotato production and proportion of farmers growing them in Matiliku Division, Makueni (Nzaui) District

Table 4 Types of sweetpotato cultivars cited and proportion of farmers growing them in Matiliku Division, Makueni (Nzaui) District

Types of sweetpotato cultivars cited	Percentage (%) of farmers growing them
Kareti	4.5
KALRO	1.5
Kirinyaga	7.6
Kitharu	1.5
KSP20	3.0
Local	21.2
Mukanda	1.5
Mulala	1.5
Munyala	1.5
Mwezi moja	10.6
Nyawo	4.5
Sallyboro	15.2
SPK 013	1.5
SPK004	3.0
Unknown	4.5
Unknown improved	16.7

13 cultivars were thinly spread among the respondents with proportions ranging between 1.5% and 7.6%. The multiplicity of cultivars suggests that the farmers have accepted sweetpotato as an important part of their food security technologies and it seems they have been searching for a suitable cultivar. It seems a combination of Sallyboro and "Mwezi moja" have the qualities the farmers have been looking for: namely, food and nutrition security. There is a lot of potential to commercialize sweetpotato production in Matiliku if a sustainable seed system is established.

According to Table 5, the most popular method of utilization of sweetpotato is boiling and taken for breakfast. The sweetpotato is also a tradable commodity with 26.5% of respondents selling them in the local market. About 6% of the late adopters had not utilized sweetpotato at all. A few farmers were feeding sweetpotato vines to livestock while there was very little processing. In the area of utilization, there is a lot of room for training and improvement. However, farmers here like in the major sweetpotato producing areas in Western and Central Kenya could target the huge fresh market in Nairobi and other major urban centers.

Table 6 shows the origin of sweetpotatoes is mainly from KALRO (35.8%), neighboring farmers (34.0), and FIPs Africa (20.8). It seems KALRO and FIPs Africa have made major inroads in addition to the farmers being receptive and a willing congregation.

Table 5 Method of utilization of sweetpotato and proportion of farmers involved in Matiliku Division, Makueni (Nzaui) District

Method of utilization	Percentage (%) of farmers utilizing them
Boil for breakfast	42.2
Boil for lunch	1.2
Chew raw	8.4
Feed vines to livestock	3.6
Have not utilized	6.0
Process into cakes	3.6
Roast	7.2
Sell in the market	26.5
Sell vines	1.2

Table 6 Origin of sweetpotato cultivars cited and proportion of farmers growing them in Matiliku Division, Makueni (Nzaui) District

Origin of sweetpotato cultivars cited	Number of times cited by (respondents) farmers	Percentage (%) of farmers growing them
Farmers	18	34.0
FIPs Africa.	11	20.8
KALRO	19	35.8
Local market	1	1.9
Unknown	4	7.5

Table 7 Education level, occupation, and main mode of communication by farmers involved in cassava and sweetpotato production in Matiliku Division, Makueni (Nzaui) District

Education level	Occupation	Telephone	Percentage (%) of farmers within the main categories
Postsecondary	Farmer	Cellphone	8
Postsecondary	Farmer/service provider	Cellphone	2
Primary	Farmer	Cellphone	34
Primary	Farmer	Cellphone/ email	2
Primary	Farmer	None	6
Primary	Farmer/ shopkeeper	Cellphone	2
Secondary	Farmer	Cellphone	36
Secondary	Farmer	Cellphone/ email	2
Secondary	Farmer	None	2
Secondary	Farmer/ shopkeeper	Cellphone	6

Results in Table 7 show that the majority, over 70%, of cassava and sweetpotato farmers had attained at least the Primary Level of education with more than half of them having a secondary level of education. At least 2% of the farmers had an email address. The results suggest that the majority of the farmers were educated and had already started embracing modern technology meaning it will be easy to convince them to adopt the new cultivars.

Cassava: Postanalysis

Cassava production and its optimized food, nutritional, and industrial positioning as a climate smart crop faces challenges including diseases, late maturing varieties, pests and lack of climate smart adaptable varieties, low yields, weak seed systems, insufficient value addition, limited market linkages, and insufficient mapping of gendered roles in production and marketing. Limited availability of clean planting materials has resulted in few agro producers growing improved varieties, hence reduced root yield and quality. The majority of cassava famers (93%) in Kenya use planting materials from their own or neighbor's fields (Githunguri et al. 2014), hence a continued build-up of diseases and pests (Githunguri and Njaimwe 2013a, b, c, d; Githunguri 1983a, b). Unfortunately, the perception of cassava as a poor peoples' food has impacted negatively on national efforts to promote cassava as a viable, commercially marketable product which has confined it to subsistence production, rudimentary processing, and limited consumption (Githunguri et al. 2017a). There is need to address and change this negative attitude towards cassava through advocacy and change in policy. The minimal processing techniques are usually tedious, time-consuming, low yielding resulting in products with unpredictable qualities and hence rarely stocked in markets. We believe that adjustments of locally available processing technologies accompanied by an aggressive vigorous promotion and marketing campaign will significantly raise the profile of climate smart crops like cassava and thus contribute to the improvement of food security, incomes, and community resilience.

Cassava (Manihot esculenta Crantz) is the fifth most important food crop in the world (Githunguri 2002). It produces more energy per unit area compared to most cereals (Githunguri et al. 2015; Githunguri and Amata 2015). The crop is drought tolerant and produces under marginally fertile soils where other crops fail (Githunguri et al. 2014); hence, it is the most resilient to climate change among all major African crops (Jarvis et al. 2012). Cassava is grown by smallholder agro producers in western, coastal, and eastern Kenya. Considering that about 75% of Kenya is arid or semi-arid and the majority of small-scale farmers in cassava growing areas are women, increased cassava production and consumption will contribute greatly towards poverty alleviation, food security, women empowerment, and wealth creation of the nation (Githunguri et al. 2007). In 2017 the country produced 1,112,000 tonnes from 90,394 ha, translating to 12 t/ha which is lower compared to 16 to 24 t/ha in China, Indonesia, and Thailand (FAO, IFAD, UNICEF, WFP and WHO 2017) and the crop potential of 90 t/ha (Cock et al. 1979). Lamu and Kisumu are among the major cassava producing counties in Kenya and the crop contributes 6% of the total household incomes (MOALF 2018). Production is dominated by low-yielding varieties that are susceptible to pests and diseases under poor crop and pest management practices. Major diseases are cassava mosaic (CMD) and cassava brown streak (CBSD) disease which cause yield losses of up to 100% with CBSD causing necrosis which renders roots unfit for food, feed, and industrial purposes (Monger et al. 2010). Pests such as cassava mealybugs (CMB) and cassava green mites (CGM) affect production particularly during the dry periods. Kenya Agricultural and Livestock Research Organization (KALRO) has identified improved varieties such as MM95/0183, Midgyera, and Katune (990005) and that are high yielding, early maturing, and tolerant to pests and diseases. Therefore, validating and upscaling improved varieties and good agronomic, pest and disease management technologies will contribute to increased productivity. Wide adoption of improved varieties will be enhanced if planting materials are bulked in close proximity to production areas to reduce transportation costs. Lack of an efficient seed delivery system has resulted in limited availability and accessibility of clean planting materials, resulting in 93% of the agro producers in Kenya using infected planting materials (Amata et al. 2012; Ekanayake and Githunguri 2000). This causes a buildup of pests and diseases further contributing to low productivity of clonally propagated crops (Githunguri et al. 1992). There is a need to have a system compliant with national regulations to avail high quality planting materials to agro producers, to address losses attributed to these pests and diseases. Use of tissue culture and three node technology to mass propagate disease free planting materials and bulk at three levels – primary (NARS, breeders' seed), secondary (Government, NGOs and other private Institutions), and tertiary (village seed entrepreneurs) – could be beneficial in availing clean materials to agro producers. Cassava utilization in the target counties is limited to boiling, deep frying, roasting, and blending cassava with maize, millet, or sorghum flour to make porridge or "ugali" (Githunguri 1995). In Nigeria, the crop is processed into feeds and food products like garri. Cassava can replace 10–30% of maize in food and feeds, 60–70% in confectionery products, and up to 60% of barley in beer making, hence can relieve the demand for and importation of wheat and save on foreign exchange. This potential has been recognized and the Ministry of Agriculture in Kenya is enacting a law to enforce blending of maize and wheat with cassava flour. Production of high quality flour is a requirement for blending and hence improved processing technologies that will ensure safety of products will be validated and up-scaled. In addition, new products will be introduced to entrepreneurs who will be linked to markets to create employment and increase income generation.

Sweetpotato: Postanalysis

Sweetpotatoes are important in the economy of poor households and are a major source of subsistence and cash income to farmers in agroclimatically disadvantaged regions and even in high potential areas of Kenya. Constraints to sweetpotato production in Kenya include the sweetpotato weevil, sweetpotato virus complex, lack of adequate disease and pest free planting materials, poor cultural practices, lack of appropriate storage and processing technologies, and poor market infrastructure. Effective control of major biotic and abiotic stresses on sweetpotatoes through selection and breeding of clones' resistance and/or tolerant to them plus the availability of clean planting material will boost food availability significantly in arid and semi-arid lands through the establishment of sweetpotato based industries in sweetpotato growing areas. The objectives of the Root and Tuber Crops Programme in the Kenya Agricultural and Livestock Research Organization (KALRO) are to develop sweetpotato varieties that are widely adapted to diverse agro-ecological zones. The varieties should also be high yielding, early bulking, and drought resistant/tolerant, resistant to major biotic and abiotic stresses and have good root quality and especially high in ß-carotene content (orange fleshed sweetpotatoes). The Kenya Agricultural and Livestock Research Organization has recognized the importance of involving farmers in their selection and breeding research programs.

Sweetpotato produces starchy tuberous roots. The crop is resistant to drought and is mainly propagated through vines. Sweetpotato can adapt to diverse climatic conditions, survive long dry spells and can be harvested and stored on a flexible time schedule, all of which qualifies it as a food security crop. Sweetpotato tuberous roots are rich in carbohydrates and are a staple food for many Africans. They are also used to manufacture alcohol. The leaves are richer in protein and minerals than the tuberous roots both qualitatively and quantitatively and are used both as vegetable and fodder. Sweetpotato has high potential if people are sensitized to their usefulness and nutritive value.

Sweetpotato is propagated also from 30 cm vines about 2/3 s of which are inserted in the ground. Both crops have a great yield potential of 20–50 t/ha of root dry weight in the tropics. This yield potential has yet to be exploited as farmers are producing an average of less than 10 t/ha (Nweke et al. 1994). Sweetpotato provides 8% or more of the minimum calories requirements of some 750 million people in the tropics, making it one of the most important energy sources in the human diet. Sweetpotato has great potential as an industrial starch production crop and hence provide employment at the production, marketing, and sales levels. Given that the demand for food and feed remains unsatisfied and the time pressure to meet the challenge is increasing rapidly, sweetpotato can help alleviate this problem.

Sweetpotatoes have the potential to contribute towards increased and sustainable food security, poverty alleviation, and wealth and employment creation. As such, the Kenyan government needs to have challenges and opportunities addressed along the sweetpotato value chain. Kenya's Economic Recovery Strategy of 2003 focuses on creating employment and wealth in the various subsectors including agriculture which contributes about 30% to the gross domestic product and provides about 70% of employment opportunities annually (MoA 2005). The vision of the government Strategy for Revitalizing Agriculture and Kenya Vision 2030 (National Economic and Social Council of Kenya (NESC) 2007) is to make the sector profitable, commercially oriented, and competitive (MoA 2005). Value addition and efficiency in production, processing, and marketing of high quality competitive sweetpotato products in Kenya and beyond will be used to enhance household incomes, employment, and food and nutrition security.

Despite its great potential as a food security and income generation crop among rural poor in marginal lands, its utilization remains low. However, its utilization can be increased with an enhanced number of recipes. Sweetpotato can accelerate development and improve livelihoods of Kenyan rural communities as has been demonstrated in China. Studies indicate sweetpotato seed, flour, crisps, starch, livestock feeds, and ethanol as the six best bet sweetpotato technologies. However, there is need to focus on sweetpotato seed. Several small-scale sweetpotato-based technologies have been initiated in a number of countries and similar models can work for Kenya (Onyango et al. 2006). What is lacking is an effective technology transfer mechanism that makes it possible for sweetpotato farmers and processors in Kenya to test, select, and adopt or adapt the best options. There is need to train farmers and processors on good sweetpotato field production and manufacturing practices and facilitate their access to microcredit and market avenues to catalyze sufficient investment in the sweetpotato industry.

A successful commercial sweetpotato seed industry could lead to increased sweetpotato field production which could in turn lead to processing of products and better markets for sweetpotato products leading to increased incomes to the rural community thus leading to a reduction in poverty levels among them. This could also lead to increased utilization of quality sweetpotato products and enhanced food security as proposed by the ERS and the SRA. Quality sweetpotato products have the potential of greatly contributing to improved health of the rural populations and enhance nutritional security among small scale farmers.

With successful interventions, there will be need for policy makers to create policy instruments to absorb surplus sweetpotato, which will create incentives for local producers. Making it mandatory for processors to use a certain percentage of sweetpotato, flour in making livestock feeds could achieve this. This would promote use of sweetpotato in addressing rural food security, industrialization, employment creation, rural urban migration, and rural youth occupation.

The sweetpotato value chain has three main components: field production, processing, and marketing. We will build strong ties with both public and private institutions engaged in research, extension, and social development in order to accomplish this linkage by the end of project. Though sweetpotato varieties that are early maturing and resistant to diseases have been developed, many farmers in the sweetpotato growing regions still grow local varieties. Many local sweetpotato cultivars take more than 6 months to mature, whereas the Kenya Agricultural and Livestock Research Organization has sweetpotato cultivars that mature between 3 and 4 months. Intensive multiplication and dissemination of the newly bred early maturing high yielding disease and pest tolerant cultivars will result in increased sweetpotato production and ensure a source of food and income for the rural households. Early maturing, high yielding disease, and pest tolerant sweetpotato cultivars developed through participatory breeding at various Kenya Agricultural and Livestock Research Organization Centres are available.

Conclusions

There are a few early cassava and sweetpotato adopters, meaning cassava and sweetpotato are still being grown as a subsistence crop and a lot needs to be done if they are going to be commercialized in Matiliku. The farmers had sufficient experience in growing the two crops albeit at subsistence level. The farmers seemed to be searching for cultivars that combined both nutritional and food security. There is need to engage more extension service providers than there are currently in order to hasten the adoption of new cultivars. There is very little processing of these crops taking place in Matiliku suggesting that there is need to carryout training and awareness creation on their role in food security and wealth creation. Sallyboro, an improved orange-fleshed sweetpotato cultivar, and "Mwezi moja," a drought tolerant local cultivar have been accepted by the farmers. There is a lot of potential to commercialize cassava and sweetpotato production in Matiliku if a sustainable seed system is established. Cassava and sweetpotato are tradable commodities in Matiliku. Farmers could target the huge fresh market in Nairobi and other major urban centers. Majority of the farmers are educated and could easily adopt new technologies.

References

Acland JD (1985) East African crops: an introduction to the production of field and plantation crops in Kenya, Tanzania and Uganda [by] J. D. Acland. Longman [for the] Food and Agriculture Organization, London, 1971. ISBN: 0582603013. 252 p

Amata R, Ateka E, Kasina M, Githunguri C, Wangai A, Miano D, Mamati E, Githiri M, Kamau J, Ndolo P, Obiero H, Muli B, Mwaniki S, Kwach J, Lelgut D, Gichangi A, Macharia I, Kagondu A, Kuyia J, Sila M, Gathara V, Kihurani A, Agili S, Momanyi V, Mbugua B, Odero B, Wekesa

N, Ochieng V, Omari J, Mbugua F, Kitisya A (2012) Constraints and opportunities of a clean seed delivery system for cassava and sweetpotato in Kenya. "Agricultural innovation system for improved productivity and competitiveness in pursuit of vision 2030". In the Proceedings of the 13th KARI Biennial Scientific Conference held in Nairobi at Kenya Agricultural Research Institute Headquarters from 22nd to 26th October 2012. pp 612–622

Ayinde A, Dipeolu O, Adebayo K, Oyewole OB, Sanni LO, Adusei J, Westby A (2004) A cost-benefit analysis of the processing of a shelf stable cassava fufu in Nigeria. In: Book of abstracts of the sixth international scientific meeting of the cassava biotechnology network. CIAT, Cali

Azogu I, Tewe O, Ezedinma C, Olomo V (2004) Cassava utilisation in domestic feed market. Root and Tuber Expansion Programme, Nigeria, 148 pp

Cock JH (1985) Cassava. New potential for a neglected crop. Westview Press, Boulder and London, 191pp

Cock JH, Franklin D, Sandoval G, Juri P (1979) The ideal cassava plant for maximum yield. Crop Sci 19:271–279

Dixon AGO, Asiedu R, Bokanga M (1994) Breeding of cassava for low cyanogenic potential problems, progress and prospects. ActaHorticulturae 375:153–161

EFDI-Technoserve (2005) Assessment of different models of cassava processing enterprises for the south and South-East of Nigeria, including the Niger Delta. Draft Final Report submitted to IITA-CEDP, March 2005

Ekanayake IJ (1998) Screening for abiotic stress resistance in root and tuber crops. IITA Res Guide 68:46

Ekanayake IJ, Githunguri CM (2000) Implications of biophysical site characteristics on growth and sustainable cassava production in the savannas of Nigeria. Niger Meteorol Soc J 2(2):33–45

Ekanayake IJ, Osiru DSO, Porto MCM (1997a) Morphology of cassava. IITA Res Guide 61:1–30

Ekanayake IJ, Osiru DSO, Porto MCM (1997b) Agronomy of cassava. IITA Res Guide 60:1–20

Ezedinma C, Patino M, Sanni L, Okechukwu R, Ilona P, Akoroda M, Dixon A (2005) Investment options in the High Quality Cassava Flour (HQCF) enterprise. Presented at the Stakeholders meeting on Strategies on sourcing high quality cassava flour. IITA, Ibadan

FAO (1990) Roots, tubers, plantains and bananas in human nutrition. FAO, Rome

FAO, IFAD, UNICEF, WFP, WHO (2017) The state of food security and nutrition in the world 2017. Building resilience for peace and food security. FAO, Rome. 232pp. ISBN 978-92-5-109888-2

Githunguri CM (1983a) Pests of cassava. In: Proceedings: Cassava production workshop, 7th–8th February 1983, Malindi, Kenya. 6p

Githunguri CM (1983b) The place of cassava in coast agriculture and relevance to the research programme. In: Proceedings: Cassava production workshop, 7th–8th February 1983, Malindi, Kenya. 24p

Githunguri CM (1995) Cassava food processing and utilization in Kenya. In: Egbe TA, Brauman A, Griffon D, Treche S (eds) Cassava food processing. CTA, ORSTOM, Wageningen, pp 119–132

Githunguri CM (2002) The influence of agro-ecological zones on growth, yield and accumulation of cyanogenic compounds in cassava. A thesis submitted in full fulfilment for the requirements for the degree of Doctor of Philosophy in Crop Physiology, Faculty of Agriculture, University of Nairobi, 195pp

Githunguri CM, Amata R (2015) Drought mitigating technologies: an overview of Cassava and Sweetpotato Production in Mukuyuni Division Makueni District in Semi-Arid Eastern Kenya. In the book "Adapting African Agriculture to Climate Change. Transforming Rural Livelihoods. The contribution of land and water management to enhanced food security and climate change adaptation and mitigation in the African continent" Edited by Walter Leal Filho, Anthony O. Esilaba, K. P. C. Rao and G. Sridhar" ISSN 1610–2010 ISSN 1610–2002 (electronic) Climate Change Management. ISBN 978–3–319-12999-0 ISBN 978–3–319-13000-2 (eBook). DOI https://doi.org/10.1007/978-3-319-13000-2. Library of Congress Control Number: 2014955614. Springer Cham Heidelberg New York Dordrecht London. © Springer International Publishing Switzerland 2015. pp 225–235

Githunguri CM, Njaimwe AN (2013a) Cassava brown streak disease (CBSD) identification and management. KARI Information brochure series 2013/86. Kenya Agricultural and Livestock Research Organization (KALRO)

Githunguri CM, Njaimwe AN (2013b) Cassava leaf spots disease identification and management. KARI Information brochure series 2013/88. Kenya Agricultural and Livestock Research Organization (KALRO)

Githunguri CM, Njaimwe AN (2013c) Cassava anthracnose disease identification and management. KARI Information brochure series 2013/93. Kenya Agricultural and Livestock Research Organization (KALRO)

Githunguri CM, Njaimwe AN (2013d) Cassava mosaic disease identification and management. KARI Information brochure series 2013/94. Kenya Agricultural and Livestock Research Organization (KALRO)

Githunguri CM, Mailu AM, Kariuki CW, Gatumbi RW (1992) An overview of cassava diseases and pests in Kenya. In: Root and tuber pest management in Kenya. Proceedings of a national workshop, 3rd–7th August 1992 Mombasa, Kenya, pp 42–48

Githunguri CM, Ekanayake IJ, Chweya JA, Dixon AGO, Imungi J (1998) The effect of different agro-ecological zones on the cyanogenic potential of six selected cassava clones. Post-harvest technology and commodity marketing. IITA, Ibadan, pp 71–76

Githunguri CM, Migwa YN, Ragwa SM, Karoki MM (2003) Cassava and sweetpotato agronomy, physiology, breeding, plant protection and product development. Root and Tuber Crops Programme in KALRO-Katumani. Paper presented at the Joint Planning meeting organized under the Eastern Province Horticulture and Traditional Food Crops Project, held at Machakos, Kenya on 5th- 7th March 2003, 5p

Githunguri CM, Karuri EG, Kinama JM, Omolo OS, Mburu JN, Ngunjiri PW, Ragwa SM, Kimani SK, Mkabili DM (2006) Sustainable productivity of the Cassava value chain: an emphasis on challenges and opportunities in processing and marketing Cassava in Kenya and beyond. A project proposal present to the Kenya Agricultural Productivity Project (KAPP) Competitive Agricultural Research Grant Fund, Research Call Ref No.KAPP05/PRC- CLFFPS –03. KAPP Secretariat 106p

Githunguri CM, Mwiti S, Migwa Y (2007a) Cyanogenic potentials of early bulking cassava planted at Katumani, a semi-arid area of Eastern Kenya. Part 2 of 4 ISSN 1023-070/2007 African Crop Science Society, Uganda, Vol. 8, pp 925 – 927

Githunguri CM, Karuri EG, Kinama JM, Omolo OS, Mburu JN, Ngunjiri PW, Ragwa SM, Mkabili DM (2007b) Sustainable productivity of the Cassava value chain: an emphasis on challenges and opportunities in processing and marketing Cassava in Kenya and beyond. In: Githunguri CM, Kwena K, Mungube E, Gatheru M (eds). KARI Katumani research centre annual report 2006. Kenya Agricultural and Livestock Research Organization (KALRO). pp. 149

Githunguri CM, Amata R, Lung'ahi EG, Musili R (2014) Cassava: a promising food security crop in Mutomo, a semi-arid food deficit district in Kitui County of Kenya. Int J Agric Resour Gov Ecol 10(3):311–323

Githunguri CM, Lung'ahi EG, Kabugu J, Musili R (2015) Cassava farming transforming livelihoods among smallholder farmers in Mutomo a Semi-Arid District in Kenya. In the book "Adapting African Agriculture to Climate Change. Transforming Rural Livelihoods. The contribution of land and water management to enhanced food security and climate change adaptation and mitigation in the African continent" Edited by Walter Leal Filho, Anthony O. Esilaba, K. P. C. Rao and G. Sridhar" ISSN 1610–2010 ISSN 1610–2002 (electronic) Climate Change Management. ISBN 978–3–319-12999-0 ISBN 978–3–319-13000-2 (eBook). DOI https://doi.org/10.1007/978-3-319-13000-2. Library of Congress Control Number: 2014955614. Springer Cham Heidelberg New York Dordrecht London. © Springer International Publishing Switzerland 2015. pp 215–224

Githunguri CM, Lung'ahi EG, Amata R, Gatheru M, Musili R, Njiru E (2016) The effect of a dry agroecological zone on selected growth and yield parameters of three elite cassava genotypes grown in Mutomo Sub-County of Kitui County in Kenya. In: Esilaba A, Githunguri C, Kimani

S, Njeru P, Lekasi J, Maina F, Muniafu M, Miriti J, Gachene C, Baaru M, Muriuki A, Nyamwaro S, Kibunja C, Mangale N (eds) Proceedings of the 27th soil science society of East Africa and the 6th African soil science society conference, held 20–25 October 2013. Soil Science Society of East Africa (SSSEA), Nakuru, pp 1100–1104

Githunguri CM, Gatheru M, Ragwa SM (2017a) Cassava production and utilization in the coastal, eastern and western regions of Kenya. In: Klein C (ed) Chapter 3 handbook on cassava: production, potential uses and recent advances. Series: plant science research and practices. Nova Science Publishers, Inc., Hauppauge. ISBN: 978-1-53610-291-8

Githunguri CM, Gatheru M, Ragwa SM (2017b) Status of cassava processing and challenges in the coastal, eastern and western regions of Kenya. In: Klein C (ed) Chapter 11 handbook on cassava: production, potential uses and recent advances. Series: plant science research and practices. Nova Science Publishers, Inc., Hauppauge. ISBN: 978-1-53610-291-8

Githunguri CM, Gatheru M, Ragwa SM (2017c) Trend in the trade of cassava products in the coastal, eastern and western regions of Kenya. In: Klein C (ed) Chapter 19 handbook on cassava: production, potential uses and recent advances. Series: plant science research and practices. Nova Science Publishers, Inc., Hauppauge. ISBN: 978-1-53610-291-8

IITA (1990a) Cassava in tropical Africa. Reference Manual IITA, Ibadan, Nigeria, 176pp

IITA (1990b) Targeting cassava breeding and selection. In: Proceedings of the 4th West and Central Africa Root Crops workshop. Lome, Togo. 12–16 December 1988. IITA Meeting Reports Series 1988/6, pp 27–30

Jarvis A, Ramirez-Villegas J, Vanessa B, Campo H, Navarro-Racines C (2012) Is cassava the answer to African climate change adaptation? Trop Plant Biol 5(1):9–29. https://doi.org/10.1007/s12042-012-9096-7

KALRO (1995) Cassava research priorities at the Kenya Agricultural Research Institute, Cassava Priority Setting Working Group. Kenya Agricultural and Livestock Research Organization (KALRO).

Lusweti CM, Kiiya W, Kute C, Laboso A, Nkonge C, Wanjekeche E, Lobeta T, Layat S, Kakuko A, Chelang E (1997) The farming systems of Sebit: In: Summary from PRA activities. Kenya Agricultural and Livestock Research Organization (KALRO), Kitale, pp 54–67

Ministry of Agriculture (MoA) (2005) Strategy for revitalizing agriculture 2004–2014. Ministry of Agriculture, Nairobi, Kenya, 23pp

Ministry of Agriculture, Livestock and Fisheries (2018) Climate risk profile for Lamu County. Kenya County Climate Risk Profile Series. The Kenya Ministry of Agriculture, Livestock and Fisheries (MoALF), Nairobi, Kenya. 24pp. https://hdl.handle.net/10568/9629

Monger W, Tomlinson J, Booonham N, Nielsen SL (2010) Development and inter-laboratory evaluation of real-time PCR assays for the detection of pospiviroids. J Virol Methods 169 (1):207–210

Nartey F (1981) Cyanogenesis in tropical feeds and foodstuffs. In: Vennesland B (ed) Cyanide in biology. Academic Press, New York, 250pp

Naskar SK, Singh DP, Srinivasan G (1989) Performance of cassava varieties in the laterite soils of Bhubaneswar and their yield stability. J Root Crops 15(1):29–31

National Economic and Social Council of Kenya (NESC) (2007) Kenya vision 2030. Office of the President, Nairobi

Nweke FI, Dixon AGO, Asadu CLA, Folayan SA (1994) Cassava variety needs of farmers and the potential for production growth in Africa. COSCA. Collaborative study of Cassava in Africa, COSCA Working Paper No. 10

Nweke FI, Spencer DSC, Lynam JK (2002) Cassava transformation. International Institute of Tropical Agriculture, Ibadan, Nigeria, 273p

Odendo M, DW Kilambya, PJ Ndolo (2001) Sweetpotato research priorities at the Kenya Agricultural Research Institute. Paper Presented at the Stakeholders Workshop for Sweetpotato Priority Setting, held at Machakos, Kenya, 27–28th September 2001, 19pp

Odongo GO (2008) Analysis of level of toxicity (cyanogenic potential) of various cassava varieties and cassava based products. A Trade Project Report in the Kenya Polytechnic University College Department of Applied Science submitted to the Kenya National Examination Council in partial fulfilment of the requirements for award of Diploma in Food Technology, 35pp

Onwueme IC (1978) The tropical tuber crops. Yams, cassava, Sweetpotato, Cocoyams. Wiley, Chichester, 234p

Onyango C, Ngunjiri PW, Oguta TJ, Wambugu SM (2006) Small-scale processing technologies for selected traditional and horticultural food crops in Kenya. Kenya Industrial Research and Development Institute, Nairobi, Kenya, 166pp

Rai M, Dhandar DG, Varma SP (1986) Variability studies of cassava varieties on growth and yield under tropical conditions. J Root Crops 12(1):25–28

Tivana LD, Bvochora J (2005) Reduction of cyanogenic potential by heap fermentation of cassava roots. Cassava Cyanide Diseases Network News, Issue No. 6 2005, 1p

6

"Conservation Agriculture," Possible Climate Change Adaptation Option in Taita Hills, Kenya

Lilian Motaroki, Gilbert Ouma and Dorcas Kalele

Contents

Abstract

The vicious cycle of food insecurity in Kenya and Africa at large is partly attributed to the high reliance on rainfed agriculture, which makes production systems vulnerable to the adverse impacts of climate change and variability. Conservation agriculture (CA) has been disseminated as a climate-smart practice that operates on three main principles to realize the multiple benefits of making

L. Motaroki (✉)
International Centre of Insect Physiology and Ecology (ICIPE), Nairobi, Kenya

G. Ouma · D. Kalele
Institute for Climate Change and Adaptation (ICCA), University of Nairobi, Nairobi, Kenya
e-mail: gouma@uonbi.ac.ke

crop production systems more resilient to climate change impacts, enhancing food security, and providing environmental services, such as carbon sequestration. As a major source of livelihood in the Taita Hills, agriculture is constrained by climate change owing to its rainfed nature. The yield and environmental and economic benefits of CA make it a suitable alternative approach to sustainable agricultural intensification, which is fundamentally different from conventional approaches based on intensive tillage and often disrupts ecosystem functions. This chapter provides the rationale for enhancing the adoption of CA in the Taita Hills by evaluating the current challenges affecting crop production, the role of CA in addressing the challenges and its potential benefits, and the barriers that must be overcome in order to promote its wide-scale adoption. A number of constraints appear to hinder the wide-scale adoption of CA in the Taita Hills, including lack of awareness, tenure-related issues, and weak policy and institutional support. Addressing these constraints will help catalyze investments for upscaling CA in the Taita Hills, with potential for replication in other parts of the country.

Keywords

Food insecurity · climate change and variability · vulnerability · tillage · conservation agriculture

Introduction

Despite being the backbone of many sub-Saharan Africa (SSA) economies, agriculture remains predominantly (95%) rainfed and therefore vulnerable to the impacts of climate change and variability (Adhikari et al. 2015). The Fifth Assessment Report of the Intergovernmental Panel on Climate Change (IPCC) presented strong evidence that surface temperatures across Africa have increased by 0.5–2 °C over the past 100 years. The report further indicates that from 1950 onward, climate change has altered the magnitude and frequency of extreme weather events. The frequency of cold days, cold nights, and frost has decreased, whereas the frequency of hot days, hot nights, and heat waves has increased (IPCC 2014). In addition, the recently released IPCC Special Report on global warming of 1.5 °C emphasizes that there are significant negative impacts at 1.5 °C temperature increase above pre-industrial levels and highlights that the observed warming in 2017 already reached 1 °C above pre-industrial levels (IPCC 2018). Since global warming of 1.5 °C is associated with extreme weather events, it poses a high risk to unique and vulnerable systems in Africa, especially agriculture, which is highly sensitive to weather and climate variables, including temperature, precipitation, and extreme weather events, such as droughts, floods, and severe storms.

The high spatial and temporal variability of rainfall, high temperatures, extreme weather events such as droughts and floods, and degradation of the sector's natural resource base (land, water, etc.) affect the agricultural production and thus food security. Even during seasons with favorable weather conditions, poor agronomic practices have been indicated to limit crop production. According to the Montpellier Report (2015), projections show an approximately 20% increase in hunger and malnutrition levels in SSA due to climate change-related impacts. Therefore, the biggest challenge for SSA remains, i.e., meeting the required food needs through sustainable intensification practices, considering the continued reliance on rainfed agriculture.

In Kenya, the key staple crops include maize, beans, and cruciferous vegetables. As an example, maize is consumed by over 90% of the total population as a staple crop (Ochieng' et al. 2017). However, crop production is primarily done by small-holder farmers who have limited resources for investing in sustainable production practices. Kiboi et al. (2019) noted that the maize productivity in Kenya is currently very low (1.0 ton ha^{-1}) relative to the attainable potential of 6–8 ton ha^{-1}. The huge gap in maize yield is primarily attributed to low soil fertility across the country's cropland resulting from poor management and nutrient mining without proper replenishment. Erratic patterns of rainfall and high frequency of droughts interacting with lack of investment in practices that conserve soil and water also contribute to low crop yields (Ochieng' et al. 2017). As the population continues to grow and climate change impacts aggravate, the situation in Kenya is likely to worsen over the coming years. The scaling up and out of practices that can support crop production even in periods of sporadic rainfall through soil water retention and increased organic matter can help break the vicious cycle.

Conservation agriculture (CA) has been disseminated as a climate-smart practice for intensifying crop production and increasing farmers' resilience to climate change (Sommer et al. 2018). The Special Report on Climate Change by the IPCC (2019) indicates that conservation agricultural practices have the potential to restore degraded land and enhance food security. However, the statistics provided by the Food and Agriculture Organization (FAO) show that the uptake of conservation agriculture in Kenya is still very low and far from the recommended 10% adoption rate, which is considered as the threshold level that can spur widespread adoption of the climate-smart agricultural technique in the country. Statistics further show that of the 5.8 million ha of arable land in Kenya, only 40,000 ha is under CA, 70% of which is by large-scale farmers (Mutuku et al. 2020).

The agriculture in Kenya is a major driver of the economy and the entire food systems. It is a major source of livelihood for the majority of Kenyan people in terms of food security, economic growth, employment creation, and foreign exchange earnings. As a major source of livelihood and well-being in the Taita Hills, agriculture remains highly sensitive and exposed to the impacts of climate change. This emphasizes the need to create an enabling environment to increase investments in sustainable agricultural practices that enhance farmers' buffer capacity while increasing productivity in the long term. The CA practice in particular has the potential to offset these impacts and make agricultural systems more resilient to a changing climate.

This chapter therefore provides the rationale for enhancing the adoption of CA in the Taita Hills by evaluating the current challenges affecting crop production, the role of CA, and its potential benefits as well as the challenges and barriers that must be overcome in order to promote its wide-scale adoption The findings of this chapter will enable policy makers to prioritize investments for scaling up the adoption of CA practice in the Taita Hills, with potential for replication in other parts of the country.

Status of Sustainable Agriculture in Kenya

A number of practices have been promoted as sustainable practices to enhance the resilience of farming systems in Kenya. These include agroforestry, soil and water conservation techniques such as terracing and irrigation, integrated soil fertility management, crop diversification, and conservation agriculture, inter alia (Kurgat et al. 2018). Despite the potential that these farming practices present for smallholder farming households and the environment, their adoption remains generally low in Kenya. The low adoption of sustainable agricultural practices is mainly attributed to barriers, such as low awareness, inadequate technical know-how, limited access to extension services, small farm sizes, and lack of capital.

Considering the smallholder nature of farming systems in Kenya, high population growth, and fragmented land holdings, the adoption of sustainable intensification farming practices is the most viable option to increase agricultural productivity (i.e., crop yield per unit area) to meet the constantly increasing food needs (FAO 2011a). It is also important to make production systems more resilient to biotic and abiotic stresses, especially those triggered by climate change, by maintaining soils in a carbon-rich state. In addition, avoiding further degradation of agricultural land and ecosystem services and rehabilitating the already-degraded agricultural land are urgent.

To achieve this, a shift is required from the current conventional tillage (CT)-based production systems of agricultural intensification as they have negative impacts on crucial natural resources, such as soil and water (Dumansky et al. 2014). The degradation of land resources is often accompanied by a decline in crop yields, hence the need for an alternative approach to agricultural production that is both ecologically sustainable and profitable (Jat et al. 2014). Another challenge for agriculture today lies in its environmental footprint and contribution to climate change. According to the IPCC (2014), agriculture contributes approximately 30% of the total greenhouse gas (GHGs) emissions comprising CO_2, N_2O, and CH_4 and is also affected by the impacts of climate change.

The sustainable intensification of the agricultural production paradigm proposed by the FAO (2011a) emphasizes the need for agricultural practices that are productive and remunerative, which enhance both the resilience of the natural resource base and environment and provide environmental services, such as carbon sequestration. This implies that the sustainable intensification of crop production should not only reduce climate change impacts on crop production but also address the causes of climate change through GHG emission reduction. Moreover, sustainable intensification practices should enhance above- and belowground biodiversity within the crop production systems to achieve better productivity. According to Jat et al. (2014), CA delivers on all of the aforementioned goals. The CA practice promotes the soil's biological activity, thus stabilizing the yields in the long term and increasing the soil's sequestration capacity.

The Concept of CA

Conservation agriculture is described as an approach to managing agro-ecosystems for improved and sustained productivity, increased profits, and food security while preserving and enhancing the natural resource base and the environment (FAO 2014a). CA is founded on three main principles, namely:

- Minimum or no tillage: This entails the minimum soil disturbance during planting, weeding, or harvesting. In special cases, seeding may be done in strips or bands, taking care to disturb less than 25% of the soil surface (FAO 2014b).
- Maintaining permanent organic mulch on the soil comprising of crop residues or cover crops.
- Diversifying the species of crops grown through sequences, rotations, or where perennial crops are involved using a balanced mix of leguminous and non-leguminous crops.

Through the three overarching principles, CA aims to conserve, improve, and make more efficient use of natural resources through integrated management of available soil, water, and biological resources (Vanlauwe et al. 2014a). This enhances environmental conservation and sustained agricultural production. CA has also been shown to reduce crop vulnerability to extreme weather events. For instance, in drought conditions, it reduces crop water requirements by up to 30%, makes better use of soil water, and facilitates deeper rooting of crops. In extremely wet conditions, CA facilitates rainwater infiltration, reducing the risk of soil erosion and downstream flooding.

The minimum soil disturbance principle of CA emphasizes the need for direct seeding, and once the soil condition has been well developed, tillage should be eliminated altogether. Soil disturbance from cultural practices such as weeding should be as minimal as possible, and a permanent or semi-permanent organic soil cover should be maintained all year round. This can be in the form of intercrops, cover crops, or mulch acquired from the residues of the previous crop. In addition, the diversification of crop rotations, association, and sequences adapted to the local environmental conditions should be practiced regularly (Njeru 2016). These help in maintaining above- and belowground biodiversity, fixing nutrients such as nitrogen into the soil, building soil organic matter, and suppressing the buildup of pests. The crop rotation principle also enhances soil fertility when legumes are included. The sequencing of crops according to seasons in the crop rotation cycle further minimizes the buildup of insect pests and diseases while optimizing nutrient use between the different types of crops.

The Benefits of CA

Impacts on Soil Quality

Soil is considered to be of good quality when it can sustain biological productivity, maintain environmental quality, and promote plant and animal health. The positive impact of CA with respect to soil quality improvement is often exhibited by an increase in soil organic carbon, good infiltration rates, high water-holding capacity, and good soil structure (Naab et al. 2017; Bamutaze et al. 2019; Mgolozeli et al. 2020). For instance, in their study, Mgolozeli et al. (2020) demonstrated the effect of CA in improving soil organic carbon particularly in the top 10-cm soil depth, promoting higher soil aggregate stability and improving the overall condition of the soil. Comparative studies of the infiltration capacity of soils under no-tillage (NT) and CT systems showed that soils in the NT system had higher total infiltration, and in the final steady state, the NT plots had an infiltration rate that was four times that of the CT plots (Fig. 1).

A conversion to conservation tillage from the conventional practices, in line with the CA principles, therefore helps improve soil structure and soil organic carbon, reduce the soil erosion risk, conserve soil moisture, reduce soil temperature fluctuations, improve soil quality, and the regulatory role it plays in the environment (Dube et al. 2012; Kakaire et al. 2015).

Impacts on Soil Carbon

Conventional tillage practices, such as the disk and chisel plough, are often associated with the significant loss of soil carbon (Fig. 2). The intensive cultivation of agricultural land has been shown to result in low amounts of soil carbon, limiting not

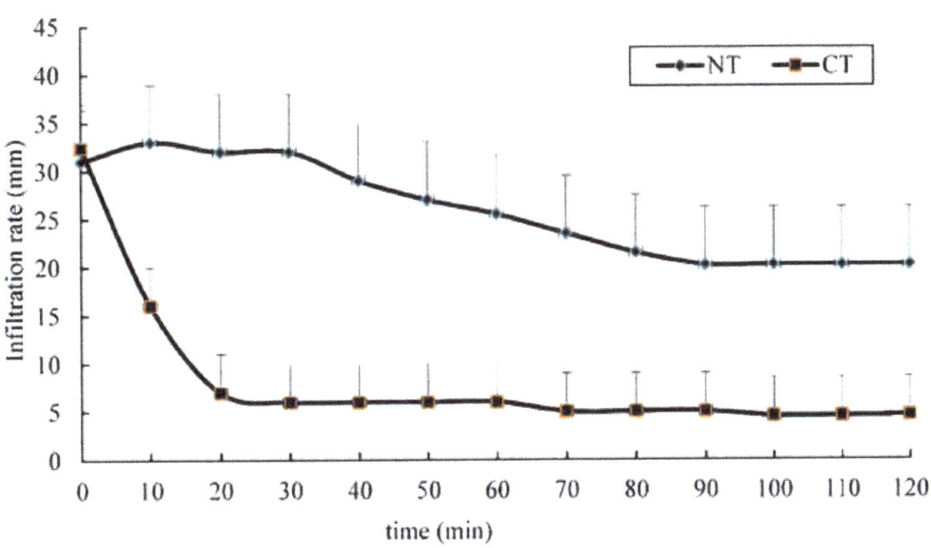

Fig. 1 Changes in the soil infiltration rate within 120 min under no-tillage (NT) and conventional tillage (CT) treatments. (Source: Mgolozeli et al. 2020)

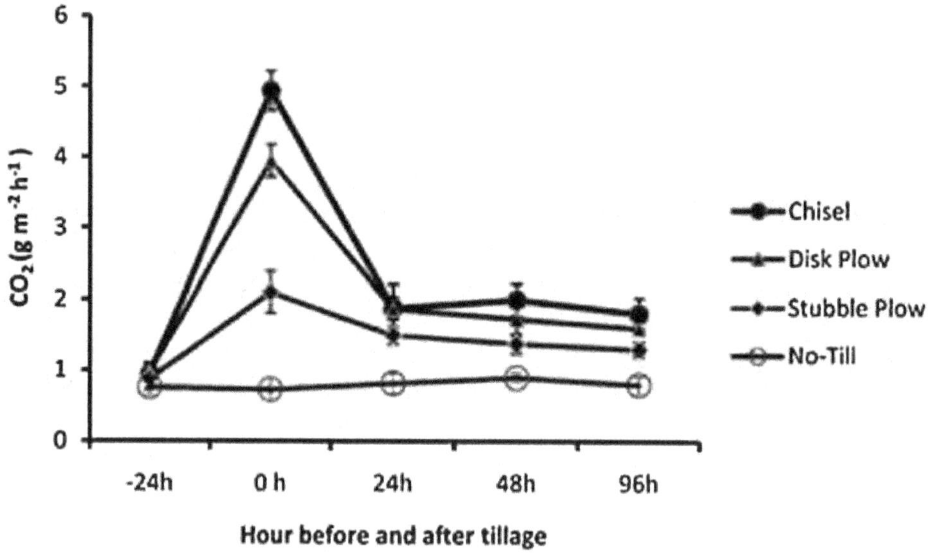

Fig. 2 CO_2 fluxes from soils associated with full tillage and NT systems. (Source: Mgolozeli et al. 2020)

just the soils' carbon sequestration capacity but also productive capacity (Dube et al. 2012). Under a changing climate, organic mulch has the potential to enhance soil productivity through the moderation of soil temperatures. For instance, combining mulching and NT can help combat climate change by enhancing carbon sequestration processes while minimizing CO_2 emissions into the atmosphere (Haddaway et al. 2016). Thus, aside from the increased productivity and soil and land restoration, CA realizes additional benefits of environmental sustainability.

Economic Benefits

In addition to the potential of CA in sustaining crop production, the practice has also been established to realize economic benefits for farmers. The findings from a study by Micheni et al. (2015) showed that although the initial costs for CA establishments, especially with weeding, were high in the first two seasons, the net income from NT systems was significantly higher compared with that of CT practices in the long term. For instance, in the fourth season, the CA costs significantly decreased to USD 24/ha compared with the constant costs of conventional systems of USD 88/ha. High cost saving was realized from the elimination of land preparation and weeding, which are common agronomic practices in CT systems. These findings are comparable with those of Guto et al. (2012) and Naab et al. (2017) who indicated that the net income from CA was higher than that of conventional systems. The low labor requirements in the CA system imply that there will be more time available to engage in off-farm income-generating activities. Additional studies from Morocco show that for many smallholder farmers, the economic gain attained from a practice supersedes the gains made in soil conservation when it comes to the adoption of labor-saving technologies (Mrabet et al. 2012). The benefits accruing from NT practices are therefore not just ecological/environmental but also reduced production costs and higher incomes to improve the well-being of farm families.

Overview of Climate Change Risks and Vulnerability in the Taita Hills

The Taita Hills are located in the Taita–Taveta County within the coastal region that lies in South Eastern Kenya. The hills form the northernmost part of the 850-km^2 stretch of the Eastern Arc Mountains in Kenya and lie within the Tsavo ecosystem which is mainly semiarid. The hills extend through different altitudinal zones depicting different agro-ecological zones within the area (Fig. 3).

Climatic and non-climatic factors interact to make the Taita Hills vulnerable to the impacts of climate change. According to Waithaka et al. (2013), some of the visible changes associated with climate change and variability in the area include the degradation of soils from moisture stress and high temperatures, limited rainfall as observed in the reduction of the volume of water in rivers, and increased frequency of extreme weather events such as floods and droughts, among others. According to the GoK (2017), the Taita Hills experience two major rainfall seasons: the long-rains season experienced between March and May (MAM) and the short-rains season experienced between October and December (OND), which in turn give rise to two planting seasons. Farmers rely on both seasons for the cultivation of their staple crops. However, the short-rains season has been noted to be more reliable than the long-rains season (GoK 2017). This is because the long-rains season is characterized

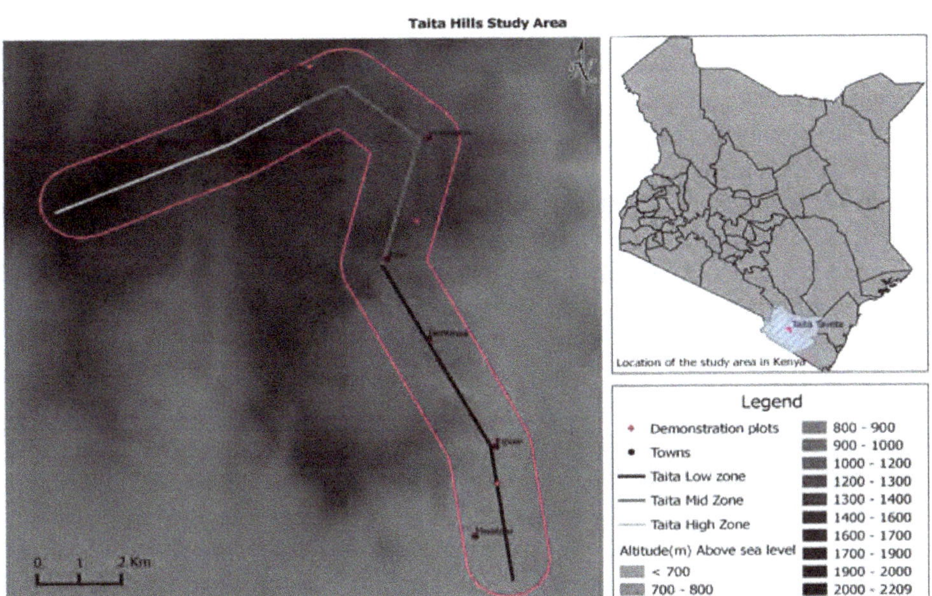

Fig. 3 Geographical location of the study area (Source: Motaroki 2016)

by, among other factors, unreliable rainfall patterns, whereby the onset of rain is delayed and cessation occurs very early in the season.

Non-climatic factors, such as high population growth, high poverty levels, poor farming methods, deforestation, and low use of technology, also have negative impacts on the status of agricultural resources, such as land and water. Due to the high rates of population growth, the area is currently characterized by small fragmented land holdings that are cultivated frequently with minimal conservation efforts, thus resulting in highly degraded soils with low fertility (Maeda 2012). A study by Muya et al. (2019) in the Taita Hills showed that in the maize-based and horticultural systems that are intensely cultivated, the amount of soil C was very low (1.6%) compared with soils under forests where the amount of C was about 7.6%. The land use intensity (LUI) in horticulture and maize-based systems was found to be over 30% compared with the LUI in forest systems (less than 2%). A similar trend was observed in the productivity index (PI), with the highest (40–50%) being recorded in natural forest and grassland and the lowest (15–20%) in horticultural and maize-based systems. The study concluded that the low soil quality and low productivity in horticulture and maize-based systems were associated with the intensification of land use as a result of inadequate knowledge on the appropriate management practices to apply. High poverty levels and lack of information cause people to engage in unsustainable practices, such as deforestation and poor farming methods, which result in land degradation. In the areas that have undergone deforestation, soil erosion has been indicated to be a major problem, with most of the terrain characterized by deep gullies.

Climatic Variability in the Taita Hills

Analysis of rainfall data for the County showed an overall declining trend in the amount of rainfall received in the various agro-ecological zones since the 1960s. The time series plot for Wundanyi (mid-altitude agro-ecological zone) showed a decline in the amount of rainfall received, the average for the years between 1993 and 2006 being less than 600 mm. This is an area that initially received between 900 and 1200 mm of rainfall per season, thus indicating a large deficit in the mentioned years (Fig. 4).

The time series analysis of Maktau in the lower zone shows the highest amount of rainfall recorded to be 626.4 mm in 1967; otherwise, there is a general declining trend in the years following 1967, with most of them receiving less than 350 mm of rainfall. This is below the average precipitation range for the lowlands which lies between 350 and 400 mm. Furthermore, the optimum rainfall condition for productive agriculture in the lowland areas is 450 mm, considering the maize variety grown. However, the time series shows the amount of rainfall received to be less than 450 mm for the period when rainfall was recorded in the weather station at Maktau (Fig. 5).

The time series analysis for the short rainy season (OND) also shows a general declining trend of observed rainfall in all the three agro-ecological zones. In the high-altitude zone, the average annual rainfall received is below 700 mm for an area that received more than 1000 mm in the past years. Similar trends are observed in the

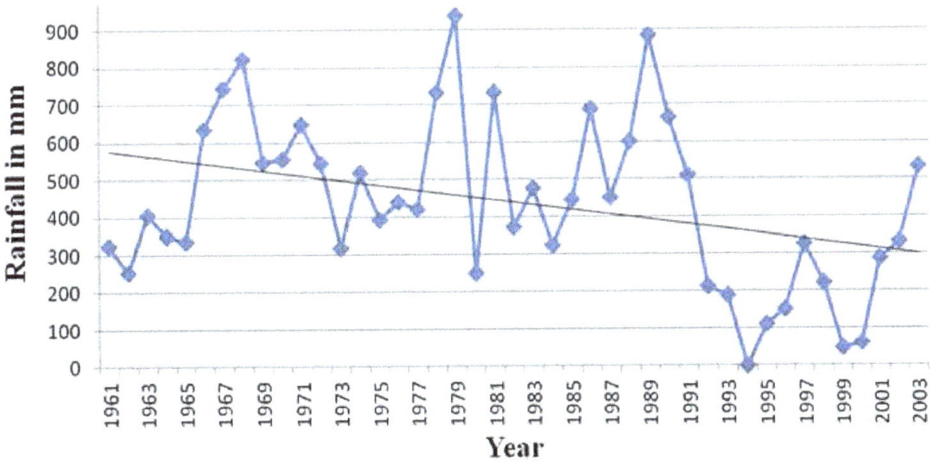

Fig. 4 Time series plot of MAM rainfall season for Wundanyi–1480 m.a.s.l (Source: Motaroki 2016)

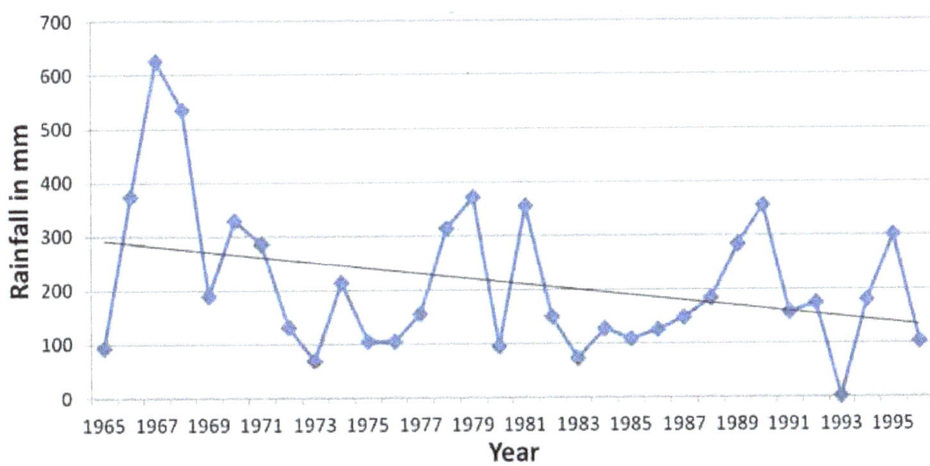

Fig. 5 Time series plot of MAM rainfall season for Maktau–700 m.a.s.l (Source: Motaroki 2016).

mid- and low-altitude agro-ecological zones where the average annual rainfall remains below 600 mm and 400 mm, respectively.

This is in agreement with the farmers' perception of variations in rainfall. The respondents from the focus group discussions (FGDs) conducted in each of the three agro-ecological zones noted that they were aware of the changes in rainfall that had taken place in the Taita Hills area over the past 20 years. The most obvious changes in rainfall reported by the respondents included drought and erratic rainfall patterns characterized by late onset and early cessation of seasonal rainfall. More specifically, drought was reported to be the most serious event, as it usually results in the loss of the entire crop, whereas erratic rainfall patterns resulted in a decline in crop yield and food shortages. Similarly, studies on the Taita Hills by Mwalusepo et al. (2015) established that the farmers were aware of the climate variability and change within their localities and the perceived changes in rainfall regimes and temperatures as well as the implications on their livelihoods.

Drought has already been indicated as a threat to food security in Kenya and is reported by the GoK (2012) as an important cause of crop losses in the Taita Hills. Observations from the IPCC (2014) show that rainfall in East Africa is likely to increase in the twenty-first century, which will be indicated by an increase in the number of extreme wet days to roughly 20%. However, even if the amount of rainfall recorded in the preceding years increased, it may not translate to improved agricultural production because the projected increases in temperature will lead to high evapotranspiration rates, thus causing water scarcity. Farmers also noted significant temperature variation with the cycles for crops like maize getting shorter and bean farming particularly in the high altitude areas that were initially very cold becoming more favorable. The high temperatures have also been accompanied by an increase in the incidence of pests and diseases that causes the decline in productivity in both crops and livestock. According to the Food and Agriculture Organization, FAO (2016), the production techniques that are part of conservation agriculture, for example, crop rotations and maintenance of a permanent soil cover, enhance soil and water conservation, thus enabling crops to withstand periods associated with rainfall variability and drought.

Crop Production Trends

The data from Osano *et al.* (2018) show that 80% of the total households in the Taita–Taveta County engage in maize production, 46% in bean crop production, and 31% in cowpea production. However, while the acreage under crop production has continuously increased, yields have continuously declined over the years due to climate change and other factors (Table 1). The overall productivity of major staple crops is very low, hence the need for better farming technologies and improved varieties. The maize crop in particular is highly sensitive to fluctuations in climate, with moisture and heat stress identified as the primary hazards for the crop.

Table 1 Short-rains crop production in the Taita–Taveta County in 2017

Crop	Acreage (ha) planted in the short-rains season	Long term average (LTA) acreage planted in the short-rains season	Actual production (90 kg bags) in the short-rains seasons	LTA production (90 kg bags) in the short-rains season
Maize	9172	9600	7849	60965
Cowpeas	805	810	4035	6830
Green grams	1079	1289	5186	5930

Source: Osano *et al.* (2018)

Table 2 Long-rains crop production in 2017

Crops	Acreage (ha) planted during the long-rains season	LTA (ha) (5 years) planted during the long-rains season	Actual long-rains production (90 kg bags)	Long-rains seasons LTA production (90 kg bags)
Maize	3681	7380	5522	22140
Beans	1069	1735	3207	10410
Green grams	1123	1566	4492	8613

Source: GoK (2017)

The land area under maize and cowpea cultivation during the 2017 short-rains season was near the long-term average (LTA), excluding the land area under green gram cultivation, which was 16% below the LTA (Osano et al. 2018). Despite the slight reduction in acreage under cultivation, the actual production of green grams was better than that of maize and beans. The 2017 short-rains seasons was also characterized by an 87% decline for maize production, 40% for bean production, and 13% for green gram production compared with the LTA (Table 1). The decline in maize production was attributed to late onset of the short rains, poor rainfall distribution both in time and in space, early cessation of rains, and crop damage caused by wildlife. Cowpea was affected by increased plucking of leaves by farmers for vegetable consumption, which negatively affected the yields and the crop's growth, hence reducing the production.

The County is mainly dependent on short rains for crop production, whereas the long rains that have increasingly become unreliable contribute 20% of the food requirements in the County. Table 2 shows the production of maize, beans, and green grams, which are the major crops grown during the long rains. The area under maize, bean, and green gram production declined by 50%, 38.4%, and 28.3%, respectively, due to prior weather forecasts, indicating the possibility of reduced rainfall during the long rains. Poor performance of the preceding short-rains seasons also contributed to the farmers' reluctance in cultivating large areas during the long-rains season (Muraguri and Nyamai 2017).

There was a significant decline in the production of all the three major crops: 75% for maize, 69.2% for beans, and 47.8% for green grams. Such a decline is mainly attributed to poor rainfall distribution in time and space as well as late onset and early cessation of rainfall. This is in agreement with the findings from a household survey targeting 600 households in the area where respondents noted that food shortages are currently on the rise, and farmers do not obtain enough yields as they did in the past. The main reasons reported for this were climatic events, including erratic rainfall patterns, above- and below-average rainfall, strong winds, floods, and frost. Other factors mentioned were high occurrence of pests, such as bean aphids, bean weevils, and cutworms, soil factors especially poor soil fertility, poor agronomic practices, and lack of inputs. The decline in crop production is a clear indication that farmers need to adopt CA which has shown potential to increase production through the integrated management of available soil, water, and biological resources

(FAO 2018). This reduces the vulnerability of crops to extreme weather events, such as droughts and floods, thus sustaining crop production.

Soil-Related Constraints

Soils are the foundation for agricultural production, hence the need for their sustainable use and management. The fertility status of soil determines the level of crop yields achieved. According to the GoK (2014), soil degradation is a major challenge to agricultural production in the Taita Hills, which leads to a decline in soil fertility. Based on the results from the household survey (Table 3), 63.3% of the respondents considered their soils to be moderately fertile, 15% noted that it was poor, 18.7% reported their soils to be very fertile, whereas 3% did not know the fertility status of their soils. When asked whether there had been a change in the fertility of their soils considering the quantity of yields obtained over the last 10 years, 55.9% indicated that it had declined, 25.2% reported it remained the same, whereas 15.2% reported that it had improved. The key factors responsible for the decline in soil fertility include continuous ploughing reported by 22.9% of the respondents, drought reported by 13.5%, use of inputs mentioned by 13.7%, floods reported by 9%, and lack of inputs reported by 2.2%. Those whose soil fertility had improved attributed such an improvement to enhanced land use practices. However, a larger percentage (29.7%) did not know the reason for the decline in the fertility of their soils.

Although the use of farm inputs, such as improved seeds, fertilizers, pesticides, and herbicides, is essential to achieve optimum crop production, assessments at the county level show that such use is generally low in the County. The low rates of input

Table 3 Soil fertility status in the Taita Hills

Soil Factors	High Zone	Mid Zone	Low Zone	Total
Soil Fertility N = 401				
Very Fertile	32	15	28	75
Moderate	89	76	89	254
Poor	20	24	16	60
Do not know	1	1	10	12
Change in Soil Fertility N = 401				
Improved	17	27	17	61
Same	37	19	45	101
Declined	86	69	69	224
No response	2	1	12	15
Reasons for the Decline in Soil Fertility				
Continuous tillage	38	28	26	92
Droughts	30	10	14	54
Use of inputs	12	40	5	57
Floods	5	11	22	38
Lack of inputs	5	3	1	9
Improved land use practices	13	7	12	32
Do not know	31	40	48	119

use is attributed to a number of reasons, including high input prices, limited access due to the distance to markets, and failure to access inputs at the right time (GoK 2014). From the survey, the various inputs used in producing maize (staple crop) include commercial fertilizer reported by 24.2% of the respondents, improved seed variety reported by 35.9%, manure reported by 30.7%, and pesticides reported by 4.5%. However, the inputs used in common bean production were minimal, including commercial fertilizer reported by only 2% of the respondents, improved seed reported by 6.2%, manure reported by 4.7%, and pesticides reported by 0.2% of the respondents. Farmers attribute their constant low yields to the use of poor-quality seed, lack of fertilizer, and manure amendments during planting.

Coping and Adaptation Strategies Currently Employed by Farmers

Just like most households in SSA, households in the Taita Hills rely on a combination of informal risk-sharing arrangements to cope with the impacts of climate change and variability. These include, for example, taking cash loans from neighbors and relatives to buy food, relying on remittances, and sending children to live with relatives (Shuaibu et al. 2014). These coping mechanisms are not sufficient or sustainable to address the challenges, and they are even more limited for the poor, landless and unemployed individuals, women, children, and large-sized households who are more vulnerable (Shuaibu et al. 2014). The respondents reported various ways in which they are coping up with the climatic events impacting their households. In coping with drought events, most of the respondents (70.9%) reported that they bought food, 6.6% sought off-farm employment, 2.8% ate different types of food, 4.7% borrowed from friends and relatives, 2.8% relied on food aid from the government, and only 2.3% changed their farming practices. However, while coping strategies such as buying food may help them in the short term, it is important to find long-term sustainable solutions, such as a shift into sustainable farming practices or livelihood diversification to off-farm activities that generate income to buffer against climate shocks (Table 4).

Table 4 Actions taken to cope with drought

Action (N=213)	High Zone	Mid Zone	Low Zone	Total
Did nothing	4	1	8	**13**
Assistance from friends/relatives	3	3	4	**10**
Relied on savings	5	2	1	**8**
Government food aid	2	3	1	**6**
Changed farming practice	0	2	3	**5**
Bought food	48	35	68	**151**
Sought off-farm employment	5	5	4	**14**
Ate different types of food	2	1	3	**6**

(Source: Motaroki 2016)

According to the MoALF (2016), the government of the Taita–Taveta County has supported and promoted adaptation strategies, such as drought-tolerant varieties of crops, rainwater-harvesting techniques, growing of trees, terracing, and other climate-smart agricultural practices. However, despite the critical role that CA plays in restoring degraded soils, stabilizing yields in the long term, and enhancing food security, few studies have attempted to assess its role in making smallholder farming systems more resilient to the impacts of climate change.

Conservation Agriculture to Enhance Adaptation in the Taita Hills

Projections show that in the period 2021–2065, the temperature in the Taita–Taveta County will increase by 0.4 °C, with greater changes expected in the long-rains season (MAM) (MoALF 2016). During the same period, projections show a 0.8% increase in precipitation in the long-rains season and a 6% increase in the short-rains season (OND). However, high evapotranspiration rates resulting from high temperatures will lead to prolonged moisture stress. The number of consecutive days experiencing moisture stress is expected to increase from 70 to around 85–90 in the long-rains season but is projected to decrease from 80 consecutive days to approximately 30 in the short-rains season.

Conservation agriculture in the Taita Hills has the potential to restore soil fertility and sustain agricultural production by stabilizing yields even during periods of sporadic rainfall and increased moisture stress, which are expected to increase in the future, while achieving other co-benefits, such as soil erosion control and reduced labor demands. However, making substantial gains from CA requires understanding when and where the practice is most effective. According to Mutuku et al. (2020), smallholder farming systems have large variations in terms of soil properties and rainfall conditions, which in turn affect the impact of CA-based practices on crop yields. There are also differences in the physical, biological, and chemical conditions of soils at the field scale as well as variations in weather patterns over time and scale.

Case Study of CA in the Taita Hills

Although few studies have tried to assess the potential of CA in the Taita Hills, the experimental assessments of CA in comparison with conventional farming practices undertaken by Motaroki (2016) established that the practice can help in increasing crop production, particularly in the semiarid lowlands.

Case Study 1

A participatory learning and action approach was used to set up demonstration plots in each of the three agro-ecological zones to compare yields from conventional tillage practices and CA for two consecutive seasons. There was no visible difference in the key yield indicators for the two seasons

(continued)

examined in both the high and mid-altitude agro-ecological zones. This could be explained by the fact that farmers in the high and mid-zones do not experience extreme conditions of drought as the lower zones. However, for the two seasons studied, there was a large visible difference in all the yield indicators for the CA treatment compared with the conventional tillage practices in the low-altitude agro-ecological zone. The zone lies in an arid and semiarid land area characterized by low rains for most parts of the year and poor soils. The effect of mulch in conserving soil moisture as well as the combined effect of organic and inorganic fertilizers could explain the immediate change in the observed yield for the area. As a matter of fact, in the second season, only the CA plot sustained crop growth, whereas the entire crop died when the conventional practices were used due to low and erratic rainfall.

Case study 1 emphasizes the importance of soil protection using mulch and the role that good agronomic practices play in minimizing the growth of weeds, promoting the growth of healthy plants, and consequently improving crop yields, especially in the drier lower zone. The conventionally grown farmers' practice plots are characterized by poor crop growth and exposure to surface runoff. Furthermore, the fact that only the CA treatment yielded results during the JJAS season further reinforces the importance of CA in arid and semiarid zones as well as the need to invest in drought-tolerant varieties of crops. Similarly, on-farm experimentation of CA by Micheni et al. (2015) showed that in drier environments, crop yield response to CA was immediate, whereas in the wetter (highland) zones, significant advantages in yield were observed from the third season onwards.

Barriers to Effective Uptake of CA

According to Njeru (2016), the issue of land tenure is a major factor influencing the adoption of CA in Kenya and many countries in Africa. There is a very low possibility of farmers investing in land with unsecured access in the long term. Particularly, considering the small sizes of land holdings in the Taita Hills, farmers have a tendency to rent in more land for their crop production. Farmers have reservations about adopting soil conservation practices such as CA because chances are high that the owners will claim their land back as soon as the fertility increases and they notice increased yields. Moreover, when farmers rent in land, they are more interested in maximizing profits from their tenancy; hence, investments in conservation practices such as CA that realizes benefits in the long term are not their priority.

In the initial years, CA increases weeds on the farm and requires additional inputs, without necessarily increasing yields. It may take up to 4 years before farmers can notice any differences in yields after adopting CA since the restoration of soil fertility takes time. In the field assessment of CA, Motaroki (2016) established that in the two seasons examined, minimal changes in yield were observed between the CA and CT plots, especially in the high and mid-altitude agro-ecological zones. As such, farmers renting a plot for less than 5 years are less likely to adopt CA as their primary concern is to increase production and profits in the short term rather than increase the quality of soils they will only use temporarily. This implies that in order to scale up and enhance widespread adoption of CA in the Taita Hills, assuring farmers of tenure security is critical.

Further, a transition from tillage-based systems to CA incurs significant costs, and for the resource-poor farmers in the Taita Hills, this may be a hindrance to their adoption of the practice. It is therefore important to provide financial support and enable the initial stages of CA adoption by giving farmers the necessary incentives in the form of subsidies. In their analysis, Kassam et al. (2014) noted that there have been cases in SSA where farmers reverted to their conventional farming practices once support for CA adoption was stopped. The above observations are in line with the results from the household surveys which showed that with regard to farmers' knowledge of the CA practice, 39.7% of the survey respondents stated that they were aware about the practice, having been introduced to it through seminar and by neighbors, agricultural extension officers, and the radio. However, only 3.7% of those who were aware about CA practiced it on their farms, stating several barriers, such as high risk and lack of economic returns (Table 5).

Cultural barriers have also been identified as one of the greatest challenges to CA adoption that has been reported in Kenya (Njeru 2016). For as long as they have been farmers, Kenyans and Africans at large have always tilled their lands such that the introduction of zero or minimum tillage faces resistance and skepticism. Another important challenge arises from the slash-and-burn system that is common in many parts of the country and Africa at large, which diminishes the availability of crop residues for mulching and incorporation into the soil (Kassam et al. 2014). Furthermore, studies show that many farmers are reluctant to use herbicides in their farms, making weed management a challenge and a hindrance to optimum crop production. Giving these farmers support in understanding new concepts and principles will create the desired change in the mindset and promote commitment to a long-term process of changing their production systems and testing and adapting new practices and technologies.

Table 5 Barriers to CA Uptake

Barrier s (N= 105)	High Zone	Mid Zone	Low Zone	Total
Small farm size	5	2	2	**9**
Expensive	1	3	4	**8**
Not profitable	29	5	10	**44**
Risk-prone	1	7	8	**16**
No specific reason	10	4	13	**27**

Studies by Pretty et al. (2011) have shown social capital to be an important prerequisite for a wide-scale adoption of sustainable practices. Farmers often have a tendency to trust their peers more than their formal advisors when evaluating potential innovations for their farms. This in effect emphasizes the importance and role of farmer organizations in embracing CA and drawing in more members within the locality to adopt the practice. Furthermore, scaling up CA to achieve impact at the local, sub-national, and national level will require a strong enabling policy and institutional environment characterized by training and enhanced access to knowledge, research, and the required inputs. If properly organized, the farmer groups or organizations can use their bargaining power to pressure local governments and institutions to create the required reforms to upscale CA (Kassam et al. 2014).

Lessons Learned, Limitations, and Recommendations for Future Research

Conservation agriculture has been shown to stabilize yields in the long term and increase the soil's ability to sequester carbon, thereby delivering on both adaptation and mitigation goals. However, a number of barriers still hinder its effective upscaling to promote widespread adoption in the Taita Hills. Overcoming these barriers will require implementing a number of measures, including incentivizing resource-poor farmers in terms of tenure security and subsidies, enhancing farmers' understanding on the benefits of CA activities such as minimum tillage, creating an enabling environment through institutional and policy support, and leveraging on social capital in target localities to promote the adoption of CA, inter alia. The case study that informed this chapter was only conducted for two seasons, which hindered the evaluation of the significant differences between the treatments used. As such, future research will need to explore the possibility of conducting field assessments over several seasons and where possible replicate the treatments in order to assess whether significant differences in the actual yield exist between CA and conventional farming methods.

Conclusion

Studies conducted in the Taita Hills have established that climate change and variability are occurring in the area as characterized by erratic rainfall patterns, high temperatures, and increased frequency of extreme weather events, such as droughts and floods. This poses a major threat to farming, which is the major source of livelihood in the area, considering that it is mostly rainfed and thus under the direct impacts of climate variability and change. Although CA has been established to maintain the soil in a carbon-rich condition and sustain production under a changing climate, its adoption remains slow due to existing barriers. To overcome these barriers and catalyze investments for upscaling and bringing about the desired reforms to advance the CA cause, strategies (policy, technological, and institutional) must be put in place for an effective adoption. These include assuring farmers of tenure security, providing financial and technical support through appropriate subsidies, capacity development, and gender-responsive policies, inter alia.

References

Adhikari U, Poyan N, Sean W (2015) Climate change and eastern Africa: A review of impacts on major crops. Food Energy Secur 4(2):110–132

Agriculture for Impact (2015) The Farms of Change: African Smallholders Responding to an Uncertain Climate Future. A Montpellier Panel Report.

Bamutaze Y, Kyamanywa S, Singh BR, Nabanoga KGN, Lal R (2019) Agriculture and ecosystem resilience in Sub Saharan Africa: livelihood pathways under changing climate. Springer, Cham

Dube E, Chiduza C, Muchaonyerwa P (2012) Conservation agriculture effects on soil organic matter on a Haplic Cambisol after four years of maize–oat and maize–grazing vetch rotations in South Africa. Soil & Tillage Research 123:21–28

Dumansky J, Reicosky DC, Peiretti RA (2014) Pioneers in soil conservation and Conservation Agriculture. Special issue, International Soil and Water Conservation Research 2(1)

FAO (2011a) State of land and water. Rome: FAO

FAO (2011b) Save and grow: a policymaker's guide to the sustainable intensifi-cation of smallholder crop production. Rome: FAO

FAO (2014a) What is Conservation Agriculture? FAO CA website http://www.fao.org/ag/ca/1a.html

FAO (2014b) CA Adoption Worldwide, FAO-CA website http://www.fao.org/ag/ca/6c.html

FAO (2016) Increased productivity and profitability of smallholder farmers through promotion and up-scaling of GAP & CA in productive semi-arid areas of Kenya (IPP-GAP). Rome. Available at http://www.fao.org/fileadmin/user_upload/FAO-countries/Kenya/docs/CA_Flyer.pdf

FAO (2018) Conservation Agriculture in Kenya: beating drought and poor harvest. Available at http://www.fao.org/fao-stories/article/en/c/1100298/

GoK (2012) Taita-Taveta District Development Plan. Government Printer.

GoK (2014) October to December 2013 short rains season assessment report. Kenya Food Security Steering Group, Nairobi, Kenya

GoK (2017) Taita Taveta County: 2017 Long Rains Food Security Assessment Report. National Drought Management Authority (NDMA), Kenya

Guto SN, Pypers P, Vanlauwe B, de Ridder N, Giller KE (2012) Socio-Ecological Niches for Minimum Tillage and Crop-Residue Retention in Continuous Maize Cropping Systems in Smallholder Farms of Central Kenya. Agronomy Journal 104(1):188–198

Haddaway NR, Hedlund K, Jackson LE et al (2016) How does tillage intensity affect soil organic carbon? A systematic review protocol. Environ Evid 5:1. https://doi.org/10.1186/s13750-016-0052-0

IPCC (2014) Summary for policymakers. In: Field CB, Barros VR, Dokken DJ, Mach KJ, Mastrandrea MD, Bilir TE, Chatterjee M, Ebi YL, Estrada YO, Genova RC, Girma B, Kissel ES, Levy AN, MacCracken S, Mastrandrea PR, White LL (eds) Climate change 2014: impacts, adaptation, and vulnerability. Part A: global and sectoral aspects. Contribution of working group ii to the fifth assessment report of the Intergovernmental Panel on Climate Change. Cambridge University Press, Cambridge, UK/New York, pp 1–32

IPCC (2018) Global warming of 1.5°C. An IPCC Special Report on the impacts of global warming of 1.5°C above pre-industrial levels and related global greenhouse gas emission pathways, in the context of strengthening the global response to the threat of climate change, sustainable development, and efforts to eradicate poverty. In: Masson-Delmotte V, Zhai P, Pörtner HO, Roberts D, Skea J, Shukla PR, Pirani A, Moufouma-Okia W, Péan C, Pidcock R, Connors S,

Matthews JBR, Chen Y, Zhou X, Gomis MI, Lonnoy E, Maycock T, Tignor M, Waterfield T (eds). In press.

IPCC (2019) Climate change and land: an IPCC special report on climate change, desertification, land degradation, sustainable land management, food security, and greenhouse gas fluxes in terrestrial ecosystems. In: Shukla PR, Skea J, Calvo Buendia E, Masson-Delmotte V, Pörtner H-O, Roberts DC, Zhai P, Slade R, Connors S, van Diemen R, Ferrat M, Haughey E, Luz S, Neogi S, Pathak M, Petzold J, Portugal Pereira J, Vyas P, Huntley E, Kissick K, Belkacemi M, Malley K (eds). In press.

Jat RA, Sahrawat KL, Kassam AH, Friedrich T (2014) Conservation agricul-ture for sustainable and resilient agriculture: global status, prospects and challenges. In: Jat R.A, Sahrawat K.L, Kassam A. H, eds. Conservation agri-culture: global prospects and challenges. Wallingford, UK: CABI, 1–25

Kakaire J, Makokha GL, Mwanjalolo M, Mensah AK, Menya E (2015) Effects of mulching on soil hydro-physical properties in Kibaale sub-catchment, South Central Uganda. Appl Ecol Environ Sci 3:127–135

Kassam A, Friedrich T, Shaxson F, Bartz H, Mello I, Kienzle J, Pretty J (2014) The spread of conservation agriculture: policy and institutional support for adoption and uptake. Field Actions Sci Rep 7:1–13

Kiboi MN, Ngetich KF, Fliessbach A, Muriuki A, Mugendi DN (2019) Soil fertility inputs and tillage influence on maize crop performance and soil water content in the Central Highlands of Kenya Agric. Water Manage 217:316–331

Kurgat BK, Ngenoh E, Bett HK, Stober S, Mwonga S, Lotze-Campen H, Rosenstock T (2018) Drivers of sustainable intensification in Kenyan rural and peri-urban vegetable production. Int J Agric Sustain 16(4–5):385–398

Maeda EE (2012) The future of environmental sustainability in the Taita Hills, Kenya: assessing potential impacts of agricultural expansion and climate change. Fennia 190(1):41–59

Mgolozeli S, Nciizah AD, Wakindiki II, Mudau FN (2020) Innovative pro-smallholder farmers' permanent mulch for better soil quality and food security under conservation agriculture. Agronomy 10(4):605

Micheni AN, Kanampiu F, Kitonyo O, Mburu DM (2015) On-farm experimentation on conservation agriculture in maize-legume based cropping systems in Kenya: water use efficiency and economic impacts. Exp Agric 52(1):1–18

MoALF (2016) Climate risk profile for Taita Taveta. Kenya county climate risk profile series. The Kenya Ministry of Agriculture, Livestock and Fisheries (MoALF), Nairobi

Motaroki L (2016) Assessment of conservation agriculture as an adaptation option to climate change in the Taita Hills, Kenya. University of Nairobi

Mrabet R, Moussadek R, Fadlaoui A, Van Ranst E (2012) Conservation agriculture in dry areas of Morocco. Field Crops Res 132:84–94

Muraguri P, Nyamai D (2017) Taita taveta County 2017: long rains food security assessment report. National Drought Management Authority, Nairobi, Kenya

Mutuku E, Roobroeck D, Vanlauwe B, Boeckx P, Cornelis M (2020) Maize production under combined conservation agriculture and integrated soil fertility management in the sub-humid and semi-arid regions of Kenya. Field Crops Research, p 254. Elsevier.

Muya EM, Karanja N, Okoth PF, Roimen H, Mung'atu J (2019) Comparative description of land use and characteristics of belowground biodiversity benchmark sites in Kenya. UoN Digital Repository, Kenya.

Mwalusepo S, Massawe ES, Affognon H, Okuku GO, Kingori S et al (2015) Smallholder farmers' perspectives on climatic variability and adaptation strategies in East Africa: the case of Mount Kilimanjaro in Tanzania, Taita and Machakos Hills in Kenya. J Earth Sci Clim Change 6:313. https://doi.org/10.4172/2157-7617.1000313

Naab JB, Mahama GY, Yahaya I, Prasad P (2017) Conservation agriculture improves soil quality, crop yield, and incomes of smallholder farmers in North Western Ghana. Front Plant Sci 8:996. https://doi.org/10.3389/fpls.2017.00996

Njeru EK (2016) Factors influencing adoption of conservation agriculture by smallholder farmers in Kenya a case of Laikipia East Sub-county Kenya. University of Nairobi, Kenya

Ochieng J, Kirimi L, Makau J (2017) Adapting to climate variability and change in rural Kenya: farmer perceptions, strategies and climate trends Nat. Resour. Forum 41:195–208

Osano K, Obondo V, Kertich J (2018) Taita Taveta County 2017: short rains food security assessment report. National Drought Management Authority, Kenya

Pretty J, Toulmin C, Williams S (2011) Sustainable intensification in African agriculture. Internat Journ of Agric Sustain 9(1):5–24

Sommer R, Paul BK, Mukalama J, Kihara J (2018) Reducing losses but failing to sequester carbon in soils – the case of Conservation Agriculture and Integrated Soil Fertility Management in the humid tropical agro-ecosystem of Western Kenya Agric. Ecosyst. Environ 254:82–91

Vanlauwe B, Wendt J, Giller KE, Corbeels M, Gerard B, Nolte C (2014a) A fourth principle is required to define Conservation Agriculture in sub-Saharan Africa: the appropriate use of fertilizer to enhance crop productivity. Field Crop Res 155:10–13

Waithaka M, Nelson GC, Thomas ST, Miriam K (2013) East African and Climate change: A comprehensive Analysis. International Food Policy Research Institute, Washington, DC

7

Agricultural Interventions to Enhance Climate Change Adaptation of Underutilized Root and Tuber Crops

Joseph P. Gweyi-Onyango, Michael Ajanja Sakha and Joyce Jefwa

Contents

J. P. Gweyi-Onyango
Department of Agricultural Science and Technology, Kenyatta University, Nairobi, Kenya

M. A. Sakha (✉) · J. Jefwa
Botany Department, National Museums of Kenya, Nairobi, Kenya

Abstract

Agricultural intensification worldwide is increasingly relying on a narrow range of crops such as rice, wheat, and maize. The reliability on this relatively small numbers of food diversities raises a very serious concern about the sustainability managing our nutrition today and in the future. We conducted a scoping review using online databases to identify various agricultural interventions that can be utilized for enhancement of underutilized root and tuber crops adaptability under the current observable effects of climate change. This is because reports of underutilized crops' adaptability to climate change continues to remain anecdotal with limited research capacity to support them. The results mooted a wide range of crop production techniques that can be utilized in production of root and tuber crops. They includes biofertilizers, tied ridging method, improved seed varieties, management of community seed banks, cropping systems, irrigation methods, exploiting abandoned lands, agroforestry practice, clean seed production technologies, and nutrient use efficiency. Based on the findings, each of these interventions plays different roles in management of the negative impacts brought up by climate change and thus they would be useful when adopted in combination since package adoption would enable farmers to benefit from the positive synergy of the selected interventions. The interventions are therefore recommended not only for sustainability but also for profitable production to meet feed, food, energy, and fiber needs and foster economic growth in the ever changing world. Therefore this chapter contributes immensely towards the development of innovative mechanisms for strengthening the resilience of root and tuber crop.

Keywords

Agricultural intensification · Sustainability · Adaptability

Introduction

The root and tuber crops are group of plants which yields tubers, starchy roots, corms, stems, and rhizomes. Okigbo (1989) defined while tuber crops as crops with edible carbohydrate-rich storage organs developing wholly or partly from underground stems while root crops as edible crops with energy-rich underground plant structures developing from modified roots. Major tropical root and tuber crops are: cassava (*Manihot esculenta*); potato (*Solanum spp.*); sweet potato (*Ipomea batata*); yam (*Dioscorea spp.*); aroids like elephant foot yam [*Amorphophallus paeoniifolius* (Dennst.) Nicolson]; taro [*Colocasia esculenta* (L.) Schott.], and tannia [*Xanthosoma sagittifolium* (L.) Schott.]. Additionally, there are minor tubers such as Chinese potato [*Plectranthus rotundifolius* (Poir.) J.K. Morton.]; yam bean [(*Pachyrhizus erosus* (L.) Urban], and arrowroot [*Maranta arundinacea* (L.)]. These crops are vegetatively propagated and plays major roles in food sector especially by managing the well-being of people in developing countries. According to FAO (2009), these crops are produced in approximately 53.93 million hectares globally and this produce about 736.747million tonnes annually.

In Sub-Saharan Africa (SSA), many people highly depend on root and tuber crops but not all as a contributory if not the primary source of their food and nutrition. This is also because of the role they play in food security, their ability to resist drought, as well as their capacity for commercial processing in Kenya. Actually they were ranked as the second most important food crop in 2019 by the ministry of agriculture after cereals (MoALF 2019). At the moment, the Kenya is producing 3.68 M MT of Irish potatoes, cassava, sweet potatoes, yam, and cocoyams. However, their yield is below average which is much way below country's potential. For instance Irish potatoes stands at 7MT per ha compared to the potential of 25 MT achieved under optimal husbandry practices (MoALF 2019).

Root and tuber crops are considered to be resilient because they are more adaptable to marginal areas. This areas are characterized by edaphic and climatic conditions that may not be favorable to the non-native materials. These crops are tolerant to poor soils and drought stress. They also grow very well on well-drained soils, with good organic matter, and especially those with loose and friable fertile clay loam or loam. Moreover, the optimal conditions for their growth are: annual rainfall ranging between 1000 and 2000 mm, temperature of 18–35 °C, and a soil pH ranging between 5 and 7.5. Nevertheless their planting differs from another case in point Coleus potato, tuberous rhizomes, or seed tubers are normally planted about 5 cm in depth on raised beds and are spaced at 15–20 cm while in Amora; they are planted at about 15 cm deep and a distance of 30–40 cm (Codd 1985).

These crops are also well known to serve as important components of subsistence farming system in their native areas and have played crucial, if not weighty roles income generation and in household food security of the rural areas. As food crops these crops are very rich in carbohydrates and on that account they play a paramount role as part of our daily diet, accounting for over 50% of the total staple food. In terms of energy requirements for global population, they contribute 3.9% energy that is sweet potato 1.5%, cassava 1.9%, yams and other root and tuber crops 0.3%. Along with this potatoes produce additional protein and dry matter per hectare than key cereals (Birch et al. 2012). Monneveux et al. (2013) also reported that potatoes have higher water generative capacity than cereals and are considered among the most energy productive crops, producing 5,600 kcal/m^3 of water, compared to 3860 in maize, 2300 in wheat, and 2000 in rice. Other root and tuber crops such as taro, yautia, and yam, also have notable energy values and inconsistent nutritional properties, including vitamin C, dietary fiber, and carotenoids (Asiedu and Sartie 2010). At the same time these crop plays a meaningful role as cash crops. They literally hold strong economic potential and can be financially rewarding to the agricultural economies. Finally these crops are progressively being used as a source of raw material for industrial use and for feeding the livestock. In comparison to other staple food crops, they provide comparatively huge amounts of nutrition and energy per unit area and time, they require less-intensive management systems even under risky environmental conditions. Consequently, these crops are very important in fighting famine caused by floods, droughts, civil strife, and other climatic catastrophe such as pests and diseases which seems to be unending in some countries.

Climate change is a menace posing extreme stress to the environment and also to the humans. Deschenes and Greenstone (2006) reported that it has an adverse effect on humans and their income-generating activities especially agriculture owing to its dependence on nature, largely temperature and precipitation. In addition climate change influences soil functions both directly and indirectly. The direct influence include soil process, namely, changes in organic matter and nutrient cycling and this is through adjusting temperature and moisture regimes or through increased soil erosion rates caused by increased frequency of intense rainfall occurrence. This causes severely effects on agriculture especially on the crops being grown. As a matter of fact, Africa is one of the riskiest continents to the ongoing climate variability causing robust negative economic impacts. This vulnerability is accentuated by development challenges specifically ecosystem degradation and endemic poverty which are supported by limited access to capital, infrastructure, markets, and technology (IPCC 2007).

Small-scale farmers in sub-Saharan Africa, who are the majority, have historically been confronted with high climate variability. Some of its negative effects on farm include decreased soil fertility and limited plant growth (Dhankher and Foyer 2018). Despite smallholder farming systems having proven to be resilient and being viable in risk-prone environments, climate change is likely to outpace their current coping capabilities (Morton 2007), if effective measures are not implemented. Specifically, low levels of income and technology, coupled with isolation from markets and lack of institutional support, are common characteristics of smallholder farming systems that make them particularly vulnerable to changes in external conditions (Morton 2007). This is worsened by the fact that food security and livelihood programs mostly stress on grain crops such as maize, rice, and wheat. In support of this, Atakos (2018) reported that only one or two studies have looked into the future potential of root tuber crops and their possible importance even with climate change. Our dependence on this relatively small number of food species therefore elevates serious concerns of feeding the whole world sustainably. In such context, investment in sustainable agricultural technologies and practices becomes crucial for adaption to sustain crop productivity to be able to feed the growing populations. In particular to able to reduce the negative effects of climate change on the agri-food system, Sombroek and Gommes (1997) proposed that populations and economic systems must be able adapt to future climatic conditions.

Since root and tuber crops play a great role as source of nutrition, and on the other hand, researchers are advocating for mitigation and adaptation as possible options to combat the adverse effects of climate change on agriculture, this chapter therefore focus on agricultural interventions that can enhance climate change adaptation of underutilized root and tuber crops.

Major Roots and Tuber Crops

Cassava

Cassava (*Manihot esculenta*) is a perennial woody shrub which grows as an annual crop. It is also referred as manioc, mandioca, or yuca, which is in the spurge family (Euphorbiaceae) (Hillocks et al. 2002). This crop is also known as the "king of tropical tuber crops" and has a significant position in the global agricultural economies. According Bennett (2015), the crop is ranked as the second most important food source in Africa with regard to calories consumed per capita. Cassava is native to South America (Allem 2002) but it is grown all over tropics and subtropics. It is largely produced in Brazil followed by Thailand, Nigeria, DR Congo, and Indonesia, even though about half of the global production is in Africa. The crop is grown in about forty African countries where it is recognized as an important food crop particularly in Nigeria, DR Congo, Ghana, Mozambique, Uganda, Cameroon, Madagascar, Angola, Côte d'Ivoire, Tanzania, Benin, and Kenya. FAO 2000 revealed that about 70% of Africa's cassava production is obtained in Nigeria, Tanzania, and DR. Congo.

It has been proposed that cassava could potentially be hardy to climate change than other staple crops (Jarvis et al. 2012). Likewise Nweke et al. (2002) revealed that cassava can grow well in marginal lands, and that it requires low farm inputs. The average cassava root yield is about 11.6 t/h worldwide (FAO 2018) which is exceedingly lower than its potential yield of 60 t/ha under better farming practices which was reported in some parts of Africa (Kintché et al. 2017). Even though FAOSTAT (2014) reported that world production of cassava storage roots improved tremendously from 176 to 277 million Mg between 2000 and 2013. Major limiting factors for cassava production are low soil fertility and pests and diseases.

Sweet Potatoes

Sweet potato (*Ipomoea batatas* (L.) Lam) is a herbaceous dicotyledonous plant belonging to Convolvulaceae family (Purseglove 1972). The plant has creeping, perennial vines and adventitious roots (Purseglove 1972). It is grown for its green leaves and storage roots which are very useful for human consumption, feeding animal, and to a certain extent, for industrial purpose (Woolfe 1992). For that reasons, according to Motsa et al. (2015), the crop plays a critical role for food security and income generation for many households. Consequently, the crop is extensively cultivated in tropical, subtropical, and frost-free temperate climatic areas of the world (Onwueme and Sinha 1991).

Sweet potato is ranked as the seventh most important food crop globally because it contributes majorly in terms of energy and nutrition (Marques 2015). The crop also matures at a very short time on marginal lands and play an important role in the economy of poor households (Nath et al. 2007). As stated by Ukom et al. (2009), the crop is important for its storage roots which can either be baked, cooked, fried, or roasted for human consumption. Its storage roots can also be processed into flour for baking bread, making noodles, as well as for alcohol production. In addition, the storage roots are very good source of vitamin A, vitamin C, vitamin B6, dietary fiber, manganese, copper, potassium, and iron (Baybutt et al. 2000). Even though the crop has a high storage root yield potential ranging between 20 and 50 t/ha (Kivuva et al. 2014), in Sub-Saharan Africa, this is yet to be realized since its production is still less than 10 t/ha (FAOSTAT 2017).

Yam

Yams (*Dioscorea spp.*) are tropical plants with large food reserve in their underground tubers and comprise of various species that originated from Southeast Asia, West Africa, East Africa, Brazil, and Guyana. The main species are *Dioscorea alata* (greater or water yam), *Dioscorea cayenensis* (yellow guinea yam), *Dioscorea esculenta* (lesser yam), and *Dioscorea rotundata* (white guinea yam) (Arnau et al. 2010), and this comprises both annual and perennial species. They cultivated all over the tropics and in some parts of subtropics and temperate areas. FAO (2000) reported that up to 95% of the world's production is realized in West Africa.

The tubers are very important and constitute stored wealth since they can be sold all-year-round by farmers because they can be stored for relatively longer period of time in comparison with other tropical fresh produce (Aidoo 2009). The tubers also provide a substantial amount of vitamins (vitamin B1 and C), potassium, and iron (Rudrappa 2013). Most essentially many of the yam species have high content of steroidal saponins which make them suitable for industrial use as corticosteroids precursors and anti-cancer bioactive compounds. Besides being staple food that is consumed by about 155 million people in the world, yams are grown as cash crop, as medicinal plant, and have high cultural value for the groups cultivating it (Coursey 1981).

Major hindrance for intensifying yam productivity is low soil fertility (both in terms of macro- and micronutrient deficiency) (O'Sullivan and Ernest 2007). This is because *Dioscorea spp.* are high-nutrient-demanding crops (Carsky et al. 2010). Therefore, yams still remains being categorized as orphan crop (Naylor et al. 2004).

Irish Potato

Irish potato (*Solanum tuberesum* L.) is indigenous to South America near the present border of Peru and Bolivia but not Ireland (Spooner et al. 2005). According to Robert and Cartwhight (2006), it belongs to the family solanaceae, and is named after Ireland country because it is associated with the Irish potato famine, also known as the Great Hunger. This was a historic famine caused by *Phytophthora infestans* which infected Irish potato crop. This crop plays a crucial role in the economy and is ranked number one non-grain food commodity (Rykaczewska 2013). Globally it is ranked third most important food crop in consumption after rice and wheat (Birch et al. 2012). It has a worldwide cultivation exceeding 19.34 million hectare in more than 158 countries, with an approximate yearly production of 364 million tons (FAOSTAT 2014). According to Tshisola (2014) in Africa, it is regarded as one of the most important food crops.

Practically the crop contains all the requisite dietary components such as protein, vitamins, carbohydrates, essential nutrients, and minerals (Sriom et al. 2017). Additionally it is a source income and employment opportunity in developing countries. Unfortunately, limiting factors to its production include short day lengths, low light intensities, high temperatures, and most importantly low soil fertility (Jones and Wendt 1994).

Cocoyams (Arrow Roots)

Cocoyam (*Colocasia esculenta* (L.) Schott.) is a member of Araceae family (Purseglove 1975) and is a subsistence and emergency food source globally (IFA 1992). It is an important starchy tuberous herbaceous perennial plant (Purseglove 1975). The crop has different varieties and commonly produced varieties are *Colocasia esculenta* (Taro) and *Xanthosoma sagittifolium* (Tannia). They also occur in all tropics and have been domesticated in most communities in Oceania, Africa, and Asia (Ramanatha et al. 2010).

The crop is cultivated for its edible cormels, corms, and leaves as well as other traditional uses (Pinto and Onwueme 2000). As food for consumption, it is essentially a source of calories obtained from underground corm and cormel (Davies et al. 2008). Their leaves which resemble spinach are nutritious and are source mineral and vitamin (Sefa-Dedeh and Kofi-Agyir 2002). Primarily, fresh cocoyam hold about 70–80% water, 20–75% carbohydrate, and 1.5–3.0% protein (Udo et al. 2005). It actually contains over 80% and 240% higher digestible crude protein than yam and cassava, respectively. Therefore this crop is high-ranking nutritionally than cassava and yam in terms of protein and other elements such as vitamin and mineral content. However, the crop still remains to be underexploited food resource (Onyeka 2014).

Root and Tuber Crop Production in Kenya

Ministry of agriculture, livestock, and fisheries of Kenya report indicate that roots and tuber crops production stagnated for a period of 3 years (2012 to 2015) with an average cultivated zone of about 240,000 ha, which gave a total production level of 3.3 million MT although these reduced to 2.4 MT in 2016 (MoALF 2019). This elicited the national government to draft a strategy (the national root and tuber crops development strategy 2019–2022) to help in upscaling their production. This was an initiative of the national government of Kenya, but it was being supported by other key stakeholders such as the European Union, and by organizations like Self Help Africa (Table 1).

Table 1 Production trends of roots and tuber crops in Kenya (2012–2016)

Commodity	Area (Ha)			Production (MT)				
Year	2012	2013	2014	2012	2013	2014	2015	2016
Irish potato	99,475	104,560	115,604	1,436,718	1,667,690	1,626,027	1,172,262	1,150,112
Sweet potato	66,971	58,509	61,067	859,549	729,645	763,643	1,232,332	697,324
Cassava	73,144	65,634	63,7265	930,922	935,089	858,461	709,926	571,845
Cocoyams	2,869	3,654	2,155	26,716	45,346	27,619	–	–
Yams	874	998	1210	10,143	13,569	20,028	–	–
Total	245,345	235,368	245,775	3,266,060	3,393,352	3,297,792	3,114,520	2,419,281

Source: MoALF 2019

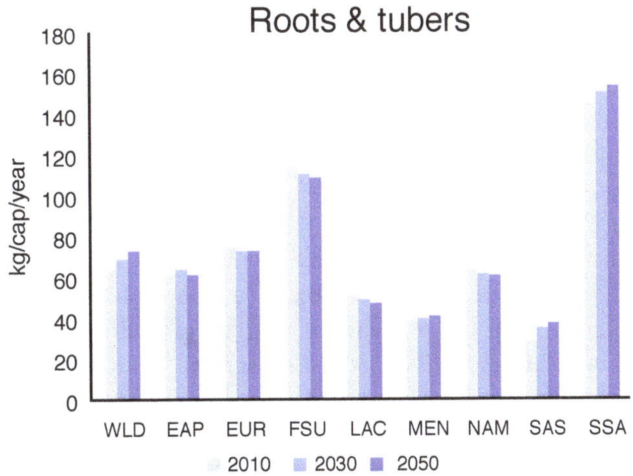

Fig. 1 Forecast sustained per capita demand for roots and tubers to 2050. (Source: Wiebe 2015). WLD = World; EAP = East Asia and Pacific; EUR = Europe; FSU = Former Soviet Union; LAC = Latin America and Caribbean; MEN = Middle East and North Africa; NAM = North America; SAS = South Asia; SSA = sub-Saharan Africa

In sub-Saharan African counties, there are noticeable cultural preferences for root and tuber crops, and using the Impact General Equilibrium Model, a 2015 analysis the International Food Policy Research Institute revealed that per capita consumption of these crops continues to rise (Fig. 1). Therefore, because of its low productivity brought by the effects of climate change and which cannot meet their demand, there is need to enhance their resilience to climate change using innovative ways.

Agricultural Interventions for Adaptation to Climate Change

Farmers, especially small-scale farmers, still use indigenous farming practices which lead up to ultimately low yields. This is coupled to farmer's nonadoption of better crop production strategies and lack of improved and high-yielding varieties. Moreover, these farmers on their own do have other alternatives that can help them bear and share loses or modifies threats. On the other hand, climate change adaptation should be built on sound and a working ecosystems, as it provides a variety of benefits and services on which agricultural production systems and rural livelihoods depend. Therefore, this calls for adoption of new technologies while producing these crop to help us throughout this challenging times. To this end, this chapter provide the suitable technologies available that can be exploited to increase root tuber crops adaptability to climate change. The technologies include:

Bio Fertilizers

By definition, bio fertilizers are products containing natural occurring microorganisms that are artificially multiplied for ameliorating soil fertility. Each and every type of crops grown in different agro-ecological zones can benefit from their use, since they are valuable to the environment. This is because they are enabling farmers to minimize the use of modern chemical fertilizers in crop production. In particular, they are designed to improve nutrient availability or reduce pest pressure. According to Malik et al. (2011), constant use of microbial-based bio-fertilizers enables microbial population to persist in the soil which helps in soil fertility conservation (Table 2).

Phosphate-Solubilizing Microorganisms (PSM): They are a group of beneficial microorganisms that capable of hydrolyzing inorganic and organic insoluble phosphorus compounds to soluble phosphorus form that can easily be absorbed by plants. They include various soil fungi and bacteria, and important species are *Pseudomonas, Bacillus, Penicillium,* and *Aspergillus.* Hikmatullah and K. Nugroho (1994) reported that their abilities of phosphate solubilization from organic materials differs due to their differences in their capacity to produce organic acids that play a key role in releasing phosphorus bound by aluminum, iron, and calcium ions. Specifically, this is because organic acids released by microorganisms are different in quantity and quality (Jha et al. 2013). Some of the organic acids released that are capable of freeing AI-P bond include Malic acid, malonic, tartaric, oxalic, and citric (Marbun et al. 2015). These organic acids decrease the soil pH in their locality to cause the dissolution of bound phosphates in soil. Khan et al. (2009) reported that 1 g of fertile soil can hold 10^1 to 10^{10} bacteria, and their live weight may exceed 2,000 kg ha^{-1}. Similarly Chen et al. (2006) outlined that among the whole microbial community in soil, phosphate solubilizing bacteria comprise 1–50% while phosphate solubilizing fungi comprise 0.1–0.5% of the total respective community. Therefore, microbial-based biofertilizers can be utilized to boost soil microbial population whenever they are low and when cultivating root and tuber crops. With high density of PSMs, it is expected that microbial phosphate solubilization can compete effectively with other microorganisms in the soil.

Phosphorus Mobilizing Fungi: This group comprise of the arbuscular mycorrhiza fungi (AM fungi) that belongs to the phylum Glomeromycota (Schüßler et al. 2001). The fungi form a mutualistic relationship with most terrestrial plants. In the association, the plants benefit through various ways such as water and essential

Table 2 Types of biofertilizers for root and tuber crops

Phosphorus-solubilizing microoganisms		
1)	Bacteria	*Bacillus megaterium* var. *phosphaticum, Bacillus subtilis, Bacillus circulans, Pseudomonas striata*
2)	Fungi	*Penicillium sp, Aspergillus awamori*
Phosphorus mobilizing fungi		
1)	Arbuscular mycorrhiza fungi	*Glomus sp., Gigaspora sp., Acaulospora sp., Scutellospora sp. & Sclerocystis sp.*
Micro nutrients solubilizers		
1)	Silicate solubilizers	*Bacillus sp.*
Plant growth promoting rhizobacteria		
1)	Pseudomonas	*Pseudomonas fluorescens*

Source: (Kumar et al. 2017)

nutrients uptake, and also by enhancing plant tolerance to biotic and abiotic stresses (Augé et al. 2015). This multifunctional ability and diversity of AM fungi has led to the development of mycorrhizal inoculants for use as biofertilizers in agriculture. Actually, most land plants are facultative symbionts such that they gain from AM fungi, yet they can also live without them, although at reasonable fitness cost. However, some plant species are obligate parasites on the fungus such that they are fully dependent on fungal nutrition (mycoheterotrophs) and have lost photosynthetic capacity (Graham et al. 2017).

Since mycorrhizal inoculations has been used for decades to stimulate plant growth for several crop, root and tuber crops which are highly mycorrhizal could also profit from the numerous services offered by this fungi not only in increasing their resilience to climate change but also their productivity. For instance, according to a report by Sieverding (1991) AM fungi inoculation of cassava increased its fresh roots weight by up to 5 t/ha.

Micronutrients Solubilizers: Such as Silicate solubilizing bacteria (SSB) are microorganisms that are able to degrade silicates and aluminum silicates in soil. These microorganisms such as Collimonas, Janthinobacterium, Proteobacteria, Aminobacter, Burkholderia, Dyella, and Frateuria are reported to solubilize the biotite which hold substantial amounts of silicate minerals (Uroz et al. 2009). During their metabolism, various organic acids are released which plays double roles in silicate weathering. These species supply hydrogen ions to the media which encourages hydrolysis. As well the organic acids let out such as oxalic acid, Keto acids, citric and hydroxyl carbolic acids form complexes with cations, which foster their removal and retention in the media in a dissolved state.

Since phosphorus and potassium are crucial macro elements for plant growth and development, P and K chemical fertilizers are regularly applied to replace the removed minerals in soil for yield optimization; SSB also plays an efficient role not only in solubilizing insoluble forms of silicates but also potassium and phosphates, hence this microorganisms when applied would increase the soil fertility, thereby enhancing root and tuber crops productivity.

Plant Growth Promoting Rhizobacteria (PGPR): Also referred as yield improving bacteria (YIB) are a group of bacteria that are known to increases plant growth and yield by-way-of several plant growth promoting substances as well as bio fertilizers. They are distinguished as free-living soil microorganisms colonizing plant roots and brings into play a beneficial effect on plant development and/or subdue plant pathogens. The microorganisms include; *Alcaligenes, Agrobacterium, Bacillus, Bradyrhizobium, Burkholderia, Enterobacter, Frankia, Klebsiella, Pseudomonas, Arthrobacter, Azospirillum, Azotobacter, Rhizobium,* and *Serratia*. The genera enclose most PGPR with well-known benefits on different crop species (Tailor and Joshi 2014).

There are numerous mechanisms through which the bacteria helps the plants including: inducement of increased nutrient uptake (termed Bio fertilizers), provision of nutrients by way of nitrogen fixation and phosphate solubilization, better plant growth promotion through the production of phytohormones (termed Bio stimulants), and the suppression of plant pathogens or the induction of systemic resistance to diseases (termed Bio protectants) (Vessey 2003). Also species of *Pseudomonas* and *Bacillus* produce plant growth regulators that stimulate crops to produce many fine roots which increases the surface area over which plant roots absorb water and nutrients.

Due to their multiple roles, research on these growth regulators has been rising and a number of experiments (both in vivo *and* in vitro) have been tested on different crops including root and tuber crops. For example, *Bacillus* and *Pseudomonas* sp. have been tried on potatoes and they have helped in improving phosphorus uptake, promoting indole acetic acid (IAA) production, and for biocontrol (Hunziker et al. 2015). Therefore, these microorganism can be very useful in fostering the growth and yield of root and tuber crops, with minor inputs of agrochemicals.

Organic Agriculture

Organic agriculture can be defined as a production system that sustains the health of soils, ecosystems, and people. This management system helps in mitigating climate change by diminishing the emissions of greenhouse gases and by sequestering carbon dioxide from the atmosphere. Organic agriculture is reported to be the most sustainable approach in food production. This is because it highly emphasizes on recycling practices and use of low external input for realization of high output. In addition, its principles dwell on increasing soil fertility, its diversity at all levels, and mitigating soil erosion.

Some management options of organic agriculture are generally considered to have both mitigation and adaptation benefits since they increase soil carbon. Ciais et al. (2013) stated that soil carbon provides a mitigation benefit by storing carbon taken out of the atmosphere by plants during photosynthesis. Soils with higher amounts of carbon are associated with greater water holding capacity, increased nutrient availability, and higher yield potentials, which could prove adaptive in a future climate (Stokes and Howden 2010). Organic agriculture practices are innovative way that can be used to increase these crops adaptability to climate change.

Soil Organic Matter Management

Soil organic matter (SOM) is a major measure of agricultural productivity and general soil health. Mean annual temperature and precipitation are major climatic drivers of SOM levels and dynamics. Long-term field trials and farm comparison show that organically managed soils have notably higher organic matter content. According to Foereid and Høgh-Jensen (2004), it was evaluated that under Northern European conditions, changing from conventional to organic agriculture resulted in enhanced SOM ranging from 100 to 400 $kgha^{-1}$ yearly during the first 50 years. Therefore, after a hundred years of organic agriculture, it is estimated that a steady state of stable level of SOM would be realized.

Environmental Protection Agency estimated that composting 1 ton of organic matter gave a net storage of about 600 pounds of carbon dioxide (EPA 2006). Even though all kinds of agriculture poses the capability of sequestering carbon, essentially organic agriculture can sequestrate remarkably more carbon than conventional systems (Don 2007). This is because organic agriculture restrains the use of chemical fertilizer and pesticide and integrates use of cover crops especially from leguminous plants, and place in order increasing soil organic matter as the first step. Therefore, to sequester more carbon as possibly brought up by climate change, it is necessary to incorporate SOM management practices when producing root and tuber crops.

Mulching

Mulching is a traditional practice which involves the covering of soil surface with organic material which plays a vital role in soil and plant protection. Mulches are capable of changing the environment around the plants and control weed sand annual grasses, soil erosion and runoff, and soil-borne diseases. Besides, they decrease moisture evaporation, increase water absorption and retention, and boost root growth. Organic or natural mulches such as compost provide many favorable and fertilizer-like effects for root and tuber crops production by supplying abundant plant nutrients, during their decaying process.

Owing to climate change that has resulted in land slide, high temperature and flashfloods, mulching is necessary since several types of mulching practices have exhibited reduction in soil erosion by more than 90% compared to bare soil (Mostaghimi et al. 1994). Unger (1994) argued that mulches with low carbon to nitrogen ratio decompose rapidly providing nutrients for crop growth at a faster rate. Furthermore, studies in Latin America and Papua New Guinea revealed the benefits of mulching cassava and sweet potato plants for yield stabilization (Ossom et al. 2001). To support this, Coling (1997) reported that mulches of plastic film enhanced **dry matter** accumulation, plant height, leaf area index, and tuber yield of potatoes. Similarly Sarma et al. (1999) after planting potato cv. Kufrimegha on ridges and flat seedbeds in combination with mulches and earthing up, he confirmed that mulching with black plastic film resulted in enhanced tuber yield which was greater than the normal cultivating method. This is because plastic mulch literally conserved the soil moisture which helped in better crop growth and tuber yield. Similarly, mulching the soil has been positively correlated with plant species richness.

As a normal practice, root and tuber crops are usually cultivated on ridges where soil erosion and weeds can be a menace. Therefore, for their effective production mulching is important since it is a valuable practice that can be used to control weeds. This is because mulching prevents weed growth and development by blocking light from reaching the soil surface where their seeds lie. In root and tuber crops, mulching could assume an important function of lowering soil temperatures in addition to soil moisture conservation (Sangakkara et al. 2004).

Zero Tillage

Zero tillage which is also referred as direct drill or no till is an agricultural practice of cultivating crops or pasture without disturbing the soil surface through tillage. Its aim is to conserve soil and moisture through nondisturbance of the soil surface and also by ensuring that 30% or more of crop residues are conserved on the surface (Erenstein and Laxmi 2008). According to Fernández et al. (2010), the practice has been documented widely for its benefits including protecting the soil against erosion

Table 3 Area under zero tillage by continent

Continent	Area (ha)	Percentage of total (%)
South America	49,579,000	46.8
North America	40,074,000	37.8
Australia and New Zealand	17,162,000	11.5
Asia	2,530,000	2.3
Europe	1,150,000	1.1
Africa	368,000	0.3
World total	115,863,000	100

Source: Derpsch et al. (2010)

and managing soil structure. Also, this practice increases carbon sequestration and activities of the microorganisms (Helgason et al. 2010) (Table 3).

Studies show that practices that reduce soil disturbance and intensify cropping have the potential of increasing soil organic matter (SOM) and improving soil health. For instance, soil disturbance with tillage generally promotes loss of SOM by facilitating microbial degradation of SOM, promoting crop-residue-soil contact, and placing residues into more favorable subsurface moisture regimes as compared to surface placement under no-tillage (Halvorson et al. 2002). By adopting no till practices in root and tuber crops production, soil conservation would be improved greatly, water and wind erosion would be considerably reduced, while the crops would yield more since they would protect from pests and external environment.

Tie-Ridging

Tie-ridges are soil and moisture conservation structures that involves the construction of small basins that are rectangular shaped. This basins are formed within the furrow of cultivated fields mainly to enhance the storage of rain water and for allowing more time for water to percolate in the soil (Wiyo et al. 1999). Belachew and Abera (2010) reported that the stored water can accessed by plants for a longer period of time better than it can be used when there is run off. Ridging across slope is highly recommended in dry areas for soil and water conservation in crop production (Kumwenda 1999).

Mechanized ridging is achieved either by animal-drawn ridgers, as it is commonly practiced in African for cassava and other food crops production, or by tractor-mounted ridgers which is a major practice for cassava production in Asian countries (Suyamto and Howeler 2001). The purpose of mechanization is to reduce hard work and in the process increase the scale of production crops. Although when using tie-ridging the optimum height of the ridge depends on the soil type and the cultivar being grown.

Benefits of tie-riding reported include: increased number of roots per plant which is identified as a major contributory factor to the higher yields on ridges (Suyamto and Howeler 2001). In particular, ridging has been shown to increase sweet potato yields by 38% (Ennin et al. 2003) over mounding, mainly as a result of increased plant population density and better weed suppression on ridges. Shetto (1999) reported that many farmers find weeding of cassava to be easier on ridges than on mounds; it is also effective in soil erosion management and results in high yields. Ennin et al. (2009) recommended the cultivation of cassava, yams, and other tuber crops on tie-ridges for realization of economizing the available planting space and thus increase on the planting density. Ridging is also responsive to improved farming practices such as herbicide application, fertilization, and yam staking, and this is due to the regular spacing of crops obtained under the systems.

The concept behind creation of tie-ridges is to optimize water infiltration, improvement of soil-water management, enhancement of root growth and nutrient uptake, and enhancement of rooting depth which is also supported by FAO (2000). Therefore, tie-ridge would play a significant role in production of root and tuber crops.

Improved Seed Varieties

Unlike the other true seed crops, root and tuber crops are propagated vegetatively. The benefits of improved seed varieties for root and tuber crops can only be realized through breeding and addressing the challenges in their seed value chain. This can be achieved through having a functioning seed systems and directly linking of the systems as a key tool of addressing the issue of improved seeds. This can further be linked in addressing the issues of climate change.

To this effect, the International potato center (CIP) has fostered and expedited their breeding schemes that shortens the time it takes in developing and releasing new varieties of root and tuber crops actually from 8 to 4 years. To this effect, their breeders have released potato varieties that are tolerant to heat, salinity, and drought. Some of the varieties are Tacna and Unica which were developed and tested in Peru; Raniag variety in Philippines; and Kinga, Kiningi, and Meva developed in Africa (Atakos et al. 2018). Further variety Tacna was released in the Republic of China under the name Jizhangshu 8, and had covered 20,000 ha by 2008. Recently work has been going on to develop climate change resilient potatoes that have the characteristics of drought and salinity tolerance, and according to Atakos et al. (2018), variety Sarnav has been released in Central Asia (Uzbekistan and Tajikistan), while heat and salinity tolerant BARI Alu-72 has been released in Bangladesh. With minimal precipitation being realized ranging between 15% and 20% and new experiences in temperature increase of 2–3 °C due to the effects of climate change, these clones have shown a high degree of tolerance to these effects (CIP 2017). On that account, such efficient breeding practices that delivers improved seed varieties of root and tuber crops should be supported especially in the area of improving their resilient to climate change.

Management of Community Seed Banks

Community seed banks are entities that are governed locally and they are managed in an informal manner by institutions whose core function is to maintain seeds for local use (Development Fund 2011). Vernooy et al. (2014) reported that the community seed banks were founded by Rural Advancement Foundation International (RAFI) now known as ETC Group or Action Group on Erosion, Technology, and Concentration. The three key functions of the community seed banks are: (i) to conserve the plant genetic resources; (ii) to make the availability and accessibility of diverse seeds and planting materials according to farmers' needs and interests easier; and (iii) facilitating seed and food sovereignty (Vernooy et al. 2014). Community seed banks are globally located in Guatemala, India, Malaysia, Mexico, Nicaragua, Sri Lanka, USA, Honduras, Bhutan, Bolivia, Brazil, Canada, Bangladesh, China, Costa Rica, Nepal, Trinidad and Tobago, Rwanda, Uganda, Mali, Burundi, Norway, Zimbabwe, and South Africa. Recently it was launched in Kenya in Nyando in smallholder farmer areas by the Consortium of International Agricultural Research Centers (CGIAR) research program on climate change agriculture and food security (CCAFS) in sub-Saharan Africa which works in partnership with Bioversity International to establish community seed banks in Kenya, Tanzania, and Uganda. The idea behind is that when you take one seed you return two, if you take five, you return ten. This protocol aids in their sustainability. The idea of community is excellent; however, the existing community seed banks are focused on cereal crops neglecting the root tuber crops. Although the International Potato Center (CIP) researchers are promoting innovative ways that allows storage of seeds of root and tuber crops.

Specifically sweet potatoes farmers are encouraged to mass-produce their own vines during planting time. The system is known as Triple S (for storing vines in sand to make them sprout). It involves storing the vines in dry sand following harvest, and thereafter planting them in seedbeds 6–8 weeks before the rainy season, and watering them to produce enough vines to plant when the rains begin. The practice results in increased vines, earlier harvests, which provides food and income at a time. In addition in order to maintain disease free planting materials, the farmers in high virus pressure zones are advised to use net tunnels systems to protect against insects such as whiteflies and aphids which are known in spreading viruses. This practice is effective in suppressing the infection rate by sweet potato virus disease thus supporting the availability of clean planting materials. According to Atakos et al. (2018), such tunnels also have contributed in moisture retention by reducing the amount of water required for irrigation. The major benefits derived from community seed banks that can be exploited for production of root and tuber crop seeds are pest and disease reduction, increased production of seeds, and climate stress buffering.

Cropping Systems

Cropping systems plays important roles in crops adaptability to climate change since the practice encompasses on farm adaptation of improved farming technologies and in this case planting more than two crops with different maturity periods. This system has various advantages including: better utilization of the environment, greater food yield, increased return per unit area, and insurance against crop failure. Therefore, root and tuber crops which have great flexibility can be part of such a system. This is because they can be intercropped with plantation crop such as areca nut, coconuts, coffee, rubber, and fruit crops like mangos, bananas, and litchi as these crops are adapted to the same ecological conditions (Nayar and Suja 2004). In this system, the principal crop provides cash while root intercrops provide as high-energy secondary staple to the farm family and feeds for farm animals, behave as insurance crop against risk and natural calamities, ensure food security, enhance resource use efficiency, augment net income, and increase employment opportunities.

To support this, cultivation of root and tuber such as yam, cassava, sweet potatoes, and edible aroids in the interspaces of perennial plantations such as rubber, coffee, coconut, and banana is common in tropical countries. Crops such as elephant foot yam and yam grow as intercrops horticultural and planation crops. Also intercropping maize and yam is believed to be productive and compatible mainly because maize is a short season crop (3–4 months), while yams are long are long duration (7–12 months) crops. Sagoe (2006) also revealed that yams and cocoyams are usually cultivated in association with cocoa and furthermore new innovations propose use of cassava trees in cocoa production. Davis et al. (1986) reported that sweet potato are grown as intercrop and in rotation systems with crops like soybean, bean, sorghum, maize, and cassava. In India, cassava, yams, and edible aroids are intercropped with rubber plantations during their immature phase whereas in Malaysia cassava is grown as intercrop in Rubber estates (Leihner 1983). The cropping system also includes practices such as adjusting the planting dates, irrigation applications, and fertilizer application (Sagoe 2006). This examples suggests that when managed appropriately, cropping systems can enhance the adaptability of these crops to climate change especially when the right crops are chosen for the system.

Irrigation Method

Climate change causes in drought which is a major abiotic factor that limit crop production. On the other hand, global warming causes rainfall fluctuations increasing the risk of plant exposure to repeated drought (Miyashita et al. 2005). Other factors ever berating the drought situation include limited fresh water bodies which cause serious issues globally, particularly in arid and semiarid regions. These fresh water bodies are also decreasing due to population pressure, comping demands from the industries, low rainfall, agricultural and urban development. Furthermore drought is considered one of the major constraints for root and tuber crop production.

Root and tuber crops have much capacity in terms of their water use efficiency and nutrient use just like other food crops. Additionally, these crops have a unique growth characteristics and functionality. Therefore a suitable irrigation method must be selected while cultivating these crops such as: an irrigation method that releases the exact amount of water required, and time and method of water application. To this end, sprinkle and trickle irrigation methods therefore becomes the best option for these crops. This is because they save irrigation water by up 30% and 50%, respectively, as compared with surface irrigation (Al-Jamal et al. 2001). Amer (2011) revealed that in small-scale farming using gated pipes and double planting row beds, with partially wetted furrow irrigation, water saving is achieved just as in trickle irrigation. Even though partially wetted irrigation can be very useful especially because it decreases installation cost compared to trickle and sprinkle systems, in this case it is important to consider the right method for climate change adaptation. Considering this, trickle irrigation becomes the best method for theses crops under the prevailing climatic conditions since it conserves water, prevents nitrogen leaching, allows deep water percolation, and reduces soil erosion. Also it gives famers an opportunity to apply water at the right time.

Exploiting Abandoned Lands

In this chapter, land abandonment concept is used on a land where traditional or recently agricultural activities have stopped due to their marginal characteristics. Lands are classified as abandoned when their productivity level is situated close to the margin beyond which management expenses are not compensated by the profits obtained after harvesting. To this effect, cropland abandonment has become a common occurrence globally due to the improper agricultural practices being used.

To reclaim this types of lands and to increase the arable lands, alternative crop productions (in intensive form) should be used. This is because the remaining arable land is being lost to urban and industrial uses, deforestation, rural development, such as afforestation, and land fragmentation into smaller units. According to FAO (2011), many efforts to revamp land productivity have continued over the years since the abandoned lands represents a valuable resource for crop production. Such efforts are supported by the farmers desire to pass productive farmlands onto the future generation in just a good condition as when they inherited them (Greenland et al. 1998).

Root and tuber crops can therefore be very useful for use in reclaiming these abandoned lands. This is because of their agronomic advantage such that these crops are adapted to diverse soil, environmental conditions, and a variety of farming methods with minimum agricultural input. Also, variations in the growth pattern and adopting cultural practices make them specific in production systems. In addition, these crops are relatively easy to grow since they are highly adaptable in growing at a wide range of altitudes, on both flat and sloping lands, and on both sandy and clay soils where other crops are not well adapted. Finally, the variety to choose from for these crops is very high.

All these characteristics indicates that root and tuber crops have a comparative advantage for cultivation in marginal lands than other crops since they can be selected to resist stress conditions and in contributing to sustainable crop production with low input cost. Similarly, IPGRI (2000) reported that these crops contributes to the food crops diversity and hence exploits fully the abilities of agro-ecosystems. Under this scenario of large abandoned land, root and tuber crop can be produced effectively on these lands, and after years of their production, the lands can be returned for production of other agricultural crops production. In return this would lead to reduced clearing of additional forests and grassland for farming purposes.

Agroforestry Practice

Agroforestry is an interaction of agriculture and trees, including the agricultural use of trees. This encompass trees on farms and in agricultural landscapes; farming in forests and along forest margins and tree-crop production including coffee and cocoa. It ranges from traditional swidden agriculture to elaborate systems of fruits trees and vines in spatial complementarity. Examples of useful agroforestry trees include; *Acacia albida*, *Leucaena leucocephala*, Prosopis juliflora, *Acacia scorpioides*, and *Euphorbia spp*. Others include green manure crops such as Tephrosia *spp,* Desmodium *spp,* Dolichos *spp*, and Indogofera *spp.*

Some of the most promising agroforestry techniques for root and tuber crops include: (a) Scattered farm trees, this simply involves the increase in number of trees in the middle of crops and alongside the farm boundaries. The trees might have greater value than the crops they displace. Alternatively, the trees might actually increase the productivity of these crops by replenishing the soil and helping reduction of soil erosion. (b) Improved fallows, which involves fallows being enriched with fast-growing trees, vines, or shrubs. This is required by the fact that the fallow periods in many areas needs to be shortened considerably. (c) Buffer strips, which are areas of land maintained in permanent vegetation that helps control soil, air, and water quality. Root and tuber crops can therefore serve as buffers against soil erosion when planted along contour lines of slopes. In complicated systems, these root and tuber crops buffer strips can include other crop, such as grasses, trees, shrubs, and fodder legumes.

Clean Seed Production Technologies

Policy makers acknowledges that seed security is pivotal for global food security. This is because seeds serves as the primary farm input in crop production and is also a means for conveying agricultural innovations to the farmers. Forasmuch as seeds are among the main factors that impediment crop production, these seeds must therefore be in good quality by the time they are distributed to the farmers. Actually availability, accessibility, and use of quality seeds that are adapted are of essence in increasing agricultural productivity and improving farmers livelihoods.

For clean seed production of root and tuber crops, cases reviewed encouraged seed stakeholders to emerge as permanent seed producers (Bentley et al. 2018). This would enable clean seed to flow through the value chain constantly. Methods of clean seed production include:

(a) Commercial farms being contracted for one season, even though the method has demonstrated that it is labor intensive, it is susceptible to pests and diseases, and it is time consuming.

(b) Tissue culture, which is the growth of tissues of plants or animal in an artificial medium separate from the parent organism. This method is very important for plant propagation for example the use of this method in seed production has ensued in mass production of clean potato seeds within a short period of time. In support of this, Pruski (2001) revealed that the method is characterized by its flexibility in terms of the rapid multiplication of seedlings. This method has been adopted in several countries like Vietnam where it revolutionized production of potato seeds. The method is now well understood and has been used in propagation of several plant species across the globe.
(c) Aeroponic method: which is the process of cultivating plants in an air or mist environment without the use of soil or media. It is considered as one of the safest and ecological friendly method for producing natural and healthy crops. The multiplication of potatoes using this method of more advantage than other available systems. CIP (2018) reported that the method is ten times more successful than convectional techniques, hydroponics, and tissue cultures which takes much time and are more labor intensive. The method utilizes nutrient solution recirculation, thus water is not wasted and the available nutrients are used effectively. Therefore this system comparatively offers lower energy and water inputs per growing area. This system can therefore be utilized for rapid production of root and tuber crops seeds.

Nutrient Use Efficiency (NUE)

Nutrient use efficiency indicates the ability of crops to take up and utilize nutrients for yield. The concept entails three major processes in plants: uptake, assimilation, and utilization of nutrients. The nutrient uptake and fertilizer recommendation of important root and tuber crops are presented in Table 4. This is an indicator that root and tuber crops have a higher NUE. This is support by Nieto, C. O. (2016) who

Table 4 Nutrient uptake and fertilizer recommendation of important root and tuber crops

Crop	Tuber yield (t/ha)	Uptake (kg/ha)			Farmyard manure (t/ha)	Recommended dose (kg/ha)		
		N	P	K		N	P	K
Cassava	30	180	22	160	12.5	100	50	100
Sweet potato	14	34	6	47	5	50	25	50
Elephant foot yam	36	122	31	176	25	100	50	150
Yams	25	163	24	127	10	80	60	80
Taro	17	119	18	157	12.5	80	25	80
Tannia	20	125	37	187	25	80	50	150
Coleus	26	106	13	107	10	60	60	100
Arrow root	24	194	31	292	10	50	25	75

Source: Mohan Kumar et al. (2000)

reported that NUE of potatoes under low nitrogen input was higher than under high nitrogen input, and higher for late cultivars than for early cultivars.

Conclusion

There is an urgent need to broaden the food basket of the Sub-Saharan Africa by supporting root and tuber crops cultivation through boosting their capacity to adapt to climate change. Even though these crops have been neglected, there is evidence that the crops have greater adaptability to extreme climatic conditions and that they are more resilient to both biotic and abiotic stresses. Therefore, the suggested innovations and/or adaptation measures will help in enhancing their adaptability to climate change and also to enable them to produce more harvestable yields where major crops have failed.

References

Aidoo R (2009) An analysis of yam consumption patterns in Ghanaian urban communities. The School of Graduate Studies, Kwame Nkrumah University of Science and Technology, Kumasi. PhD Thesis. Access date 03 July 2017

Al-Jamal MS, Ball S, Sammis TW (2001) Comparison of sprinkler, trickle and furrow irrigation efficiencies for onion production. Agric Water Manag 46(3):253–266

Allem C (2002) The origins and taxonomy of cassava. In: Hillocks RJ, Thresh JM, Bellotti AC (eds) Cassava: biology, production and utilization. CABI Publishing, Wallingford, pp 1–16

Amer H (2011) Effect of irrigation method and quantity on squash yield and quality. Agric Water Manag 98:1197–1206

Arnau G, Abraham K, Sheela M, Chair H, Sartie A, Asiedu R (2010) Yams, in root and tuber crops. In: Bradshaw J (ed) Handbook of plant breeding, vol 7. Springer Science + Business Media, New York, pp 127–148

Asiedu R, Sartie A (2010) Crops that feed the world 1. Yams. Food Secur 2:305–315

Atakos V (2018) Next generation technologies: tackling climate change in agriculture. Retrieved from: https://cipotato.org/blog/next-generation-technologiestackling-climate-change-in-agriculture/

Atakos V, Thiele G, Campos H, Friedmann M, Holmes H, Dudenhoefer D (2018) Next generation technologies: tackling climate change in agriculture. CIP. December 5, 2018

Augé M, Toler D, Saxton M (2015) Arbuscular mycorrhizal symbiosis alters stomatal conductance of host plants more under drought than under amply watered conditions: a meta-analysis. Mycorrhiza 25:13–24

Baybutt C, Hu L, Molteni A (2000) Vitamin A deficiency injures lung and liver parenchyma and impairs function of rat type II pneumocytes. J Nutr 130(5):1159–1165

Belachew T, Abera Y (2010) Response of maize (*Zea mays* L.) to tied ridges and planting methods at Goro, South eastern Ethiopia. Am Eur J Agron 3:21–24

Bennett B (2015) Smallholder cassava production and the cassava processing sector in Africa. Food Chain 5:1–3

Bentley JW, Andrade-Piedra J, Demo P, Dzomeku B, Jacobsen K, Kikulwe E, Ogero K (2018) Understanding root, tuber, and banana seed systems and coordination breakdown: a multi-stakeholder framework. J Crop Improv 32(5):599–621

Birch J, Bryan G, Fenton B, Gilroy E, Hein I, Jones T, Prashar A, Taylor A, Torrance L, Toth K (2012) Crops that feed the world. Potato: are the trends of increased global production sustainable? Food Secur 4:477–508

Carsky J, Asiedu R, Cornet D (2010) Review of soil fertility management for yam-based systems in West Africa. Afr J Root Tuber Crops 8:1–17

Chen P, Rekha D, Arun B, Shen T, Lai A, Young C (2006) Phosphate solubilizing bacteria from subtropical soil and their tricalcium phosphate solubilizing abilities. Appl Soil Ecol 34(1):33–41

Ciais P, Sabine C, Bala G, Bopp L, Brovkin V, Canadell J, Chhabra A, DeFries R, Galloway J, Heimann M, Jones C, Quéré L, Myneni B, Piao S, Thornton P (2013) Carbon and other biogeochemical cycles. In: Stocker TF, Qin D, Plattner G-K, Tignor M, Allen SK, Boschung J, Nauels A, Xia Y, Bex V, Midgley PM (eds) Climate change 2013. The physical science basis. Contribution of working group I to the fifth assessment report of the intergovernmental panel on climate change. Cambridge University Press, Cambridge, UK/New York, pp 465–570

CIP (2017) Seed potato for Africa, [Accessed on April 12 2017] and available at http://cipotato.org/seed-potato-for-africa/

CIP (2018) Potato minituber boom in Sub-Saharan Africa: tenfold increase in ten years – International Potato Center

Codd E (1985) Lamiaceae (Labiatae). In: Leistner OA (ed) Flora of Southern Africa. Volume 28, Part 4. Botanical Research Institute, Pretoria. 247pp

Coling C (1997) Effect of plastic film mulching on increasing potato yield: effect of plastic film mulching on increasing potato yield. Acta Agric Zhojangensis 9:83–86

Coursey G (1981) The interactions of yam and man. J Agric Tradit Bot Appl 1:5–21

Davies M, Labuschagne T, Koen E, Benesi M, Saka K (2008) Some properties of starches from cocoyam in Malawi. Afr J Food Sci 2:102–111

Davis C, Wolley N, Moreno A (1986) Multiple cropping with legumes starchy roots. Macmillan Publishing, New York, pp 133–160

Derpsch R, Friedrich T, Kassam A, Hongwen L (2010) Current status of adoption of no-till farming in the world and some of its main benefits. Int J Agric Biol Eng 3(1):1–25

Deschenes O, Greenstone M (2006) The economic impacts of climate change: evidence from agricultural output and random fluctuations in weather. Am Econ Rev 97:354–385

Development Fund (2011) Banking for the future: savings, security and seeds. Development Fund, Oslo. Accessed 29 Jan 2016

Dhankher OP, Foyer CH (2018) Climate resilient crops for improving global food security and safety. Plant Cell Environ 41(5):877–884

Don C (2007) No shortcuts in checking soil health. Agric Res. July 2007, at 4. Available at http://www.ars.usda.gov/is/AR/archive/jul07/soil0707.pdf. Last visited 13 Nov 2008

Ennin A, Dapaah K, Asafu-Agyei N (2003) Land preparation for increased sweet potato production in Ghana. Paper presented at the 13th symposium of the international society for tropical root crops (ISTRC-World Branch). Held from 10th–14th November, 2003, Arusha, 14pp

Ennin SA, Otoo E, Tetteh FM (2009) Ridging, a mechanized alternative to mounding for yam and cassavaproduction. West Afr J Appl Ecol 15:1–8

EPA (2006) Solid waste management and greenhouse gases: a life-cycle assessment of emissions and sinks. 55

Erenstein O, Laxmi V (2008) Zero tillage impacts in India's rice–wheat systems: a review. Soil Tillage Res 100:1–14

FAO (2000) The state of food insecurity in the world 2000. ISBN 92-5-104479-1 Job No. X8200/E. 36pp

FAO (2009) FAOSTAT statistical database. Rome (available at http://faostat.fao.org)

FAO (2011) The state of the world's land and water resources for food and agriculture (SOLAW). Managing systems at risk. FAO, Rome

FAO. FAOSTAT (2018) Food and Agriculture Organization of the United Nations (FAO), Rome. http://www.fao.org/faostat/

FAOSTAT (2017) Food and Agriculture Organization of the United Nations. Rome. http://apps.fao.org

FAOSTAT (Food and Agriculture Organization Data of Statistics) (2014) Food and Agriculture Organization Data of Statistics. One hundred fifty eight countries data base. http://faostat.fao.org/site/567/

Fernández R, Quiroga A, Zorati C, Noellemeyer E (2010) Carbon contents and respiration rates of aggregate size fractions under no-till and conventional tillage. Soil Tillage Res 109:103–109

Foereid B, Høgh-Jensen H (2004) Carbon sequestration potential of organic agriculture in northern Europe – a modelling approach. Nutrient Cycling in Agroecosystems 68(1):13–24

Graham W, Lam Y, Merckx V (2017) Plastomes on the edge: the evolutionary breakdown of mycoheterotroph plastid genomes. New Phytol 214:48–55

Greenland J, Gregory J, Nye H (1998) Land resources: on the edge of the Malthusian precipice. CAB International, Wallingford

Halvorson D, Weinhold J, Black L (2002) Tillage, nitrogen, and cropping system effects on soil carbon sequestration. Soil Sci Soc Am J 66:906–912

Helgason L, Walley L, Germida J (2010) No-till soil management increases microbial biomass and alters community profiles in soil aggregates. Appl Soil Ecol 46:390–397

Hillocks J, Thresh M, Bellotti C (2002) Cassava biology, production and utilization. CABI Publishing, Wallingford, p 2002

Hunziker L, Bönisch D, Groenhagen U, Bailly A, Schulz S, Weisskopf L (2015) *Pseudomonas* strains naturally associated with potato plants produce volatiles with high potential for inhibition of *phytophthora infestans*. Appl Environ Microbiol 81:821–830

IFA (1992) World fertilizer use manual. Publication of International Fertilizer Industry Association, Paris, pp 148–160

Intergovernmental Panel on Climate Change (2007) Climate change. In: Impacts, adaptation and vulnerability. Contribution of working group II to the fourth assessment report of the intergovernmental panel on climate change. Cambridge University Press, Cambridge, UK. 976pp

International Plant Genetic Resources Institute, (IPGRI) (2000) Enhancing the contribution of neglected and under-utilized species to food security, and to incomes of the rural poor. IPGRI, Rome

Jarvis A, Ramirez-Villegas J, Campo H, Navarro-Racines C (2012) Is cassava the answer to African climate change adaptation? Trop Plant Biol 5:9–29

Jha A, Saxena J, Sharma V (2013) Investigation on phosphate solubilization potential of agricultural soil bacteria as affected by different phosphorus sources, temperature, salt and pH. Commun Soil Sci Plant Anal 44:2443–2458

Jones B, Wendt W (1994) Contribution of soil fertility research to improved maize production by smallholders in eastern and southern Africa. In: Proceedings of the fourth Eastern and Southern Africa regional maize conference 2–14

Khan MS, Zaidi A, Wani PA (2009) Role of phosphate solubilizing microorganisms in sustainable agriculture-a review. In: Sustainable agriculture (pp. 551–570). Springer, Dordrecht

Kintché K, Hauser S, Mahungu M, Ndonda A, Lukombo S, Nhamo N (2017) Cassava yield loss in farmer fields was mainly caused by low soil fertility and suboptimal management practices in two provinces of the Democratic Republic of Congo. Eur J Agron 89:107–123

Kivuva M, Yencho G, Sibiya J, Musembi G (2014) Assessment of production constraints and farmers' preferences for sweet potato genotypes. J Plant Breed Genet 02(01):15–29

Kumar A, Maurya BR, Raghuwanshi R, Meena VS, Islam MT (2017) Co-inoculation with Enterobacter and Rhizobacteria on yield and nutrient uptake by wheat (*Triticum aestivum* L.) in the alluvial soil under indo-gangetic plain of India. J Plant Growth Regul 36(3):608–617

Kumwenda F (1999) The role of animal traction in soil water conservation tillage practices among smallholder farmers in Malawi. In: Kaumbutho PG, Simalenga TE (eds) Conservation tillage with animal traction. A resource book of the Animal Traction Network for Eastern and Southern Africa (ATNESA), Harare. 173pp

Leihner D (1983) Management and evaluation of intercropping systems with cassava. CIAT, Cali

Malik A, Sanghmitra S, Javed A (2011) Chemical vs. organic cultivation of medicinal and aromatic plants: the choice is clear. Int J Med Arom Plants 1(1):5–13

Marbun S, Sembiring M, Bintang (2015) Phosphate solubilizing microbe and organic matter application to increase P uptake and potatos growth at andisol impacted Sinabung Mountain Eruption. J Agroekoteknol 4:1651–1658. (In Indonesian)

Marques M (2015) Bacterial endophytes of sweet potato tuberous roots affected by the plant genotype and growth stage. J Appl Microbiol 96:273–281

Miyashita K, Tanakamaru S, Maitani T, Kimura K (2005) Recovery responses of photosynthesis, transpiration and stomatal conductance in kidney bean following drought stress. Environ Exp Bot 53:205–214

MoALF (2019) National root and tuber crops development strategy 2019–2022. Retrieved 12 November 2020, from https://www.kilimo.go.ke/wp-content/uploads/2019/10/Roots-and-Tuber-Crops-Strategy-2019-2022.pdf

Mohan Kumar CR, Nair M, George J, Ravindran S, Ravi V (2000) Production technology of tuber crops. Central Tuber Crops Research Institute, Thiruvananthapuram, p 174

Monneveux P, Ramírez A, Pino T (2013) Drought tolerance in potato (*S. tuberosum* L.): can we learn from drought tolerance research in cereals? Plant Sci 205:76–86

Morton F (2007) The impact of climate change on smallholder and subsistence agriculture. Proc Natl Acad Sci USA 104(50):19680–19685

Mostaghimi S, Gidley M, Dillaha TA, Cooke A (1994) Effectiveness of different approaches for controlling sediment and nutrient losses from eroded land. J Soil Water Conserv 49(6):615–620

Motsa NM, Modi T, Mabhaudhi T (2015) Sweet Potato (*ipomoea Batatas* L.) As a Drought Tolerant and Food Security Crop. S Afr J Sci 111(11/12):8. Art. #2014–0252

Nath R, Chattopadhyay A, Kundu CK, Majumder A, Islam Sk J, Gunri S (2007) Production potential of sweet potato in the Red and Laterite ecosystem of West Bengal. J Root Crops 33:97–103

Nayar R, Suja G (2004) Production potential of root and tubers in multiple cropping systems involving plantation crops. J Root Crops 30(2):93–100

Naylor L, Falcon P, Goodman M, Jahn M, Sengooba T, Tefera H (2004) Biotechnology in the developing world: a case for increased investments in orphan crops. Food Policy 29:15–44

Nieto CO (2016) Nitrogen use efficiency in potato: an integrated agronomic, physiological and genetic approach. Doctoral dissertation, Wageningen University

Nugroho W (1994) Klasifikasidan Pemetaan Semi Detail Tanah Berkembangdari Abu Volkanik-berdasarkan Sistem Taksonomi Tanah di Desa Candikuning, Baturiti, Tabanan. Skripsi. Jurusan Tanah, Fakultas Pertanian Universitas Udayana. pp 75

Nweke FI, Spencer DSC, Lynam JK (2002) The cassava transformation: Africa's best kept secret, Michigan State University Press, East Lansing, USA. pp 273

O'Sullivan N, Ernest J (2007) Nutrient deficiencies in lesser yam (*Dioscorea esculenta*) characterized using constant-water table sand culture. J Plant Nutr Soil Sci 170(2):273–282

Okigbo N (1989) New crops for food industry: the roots and tuber in tropical Africa. In: New crops for food industry. G.E. Weekens et al, Chapman and Wall/London, pp 123–134

Onwueme C, Sinha D (1991) Field crop production in tropical Africa: principles and practice. Technical Centre for Agriculture and Rural Cooperation, Ede. ISBN-13: 9789290810865, pp 267–275

Onyeka J (2014) Status of cocoyam (Colocasia esculenta and Xanthosoma spp.) in west and central Africa: production, household importance and the threat from leaf blight. CGIAR Research Program on Roots, Tubers and Bananas (RTB)

Ossom M, Pace F, Rhykerd L, Rhykerd L (2001) Effect of mulch on weed infestation, soil temperature, nutrient concentration and tuber yield in *Ipomoea batatas* in Papua New Guinea. Trop Agric 78:144–151

Pinto J, Onwueme B (2000) Food values of breadfruit, taro leaves, coconut and sugar cane. FAO. Plant production and protection paper 126. FAO, Rome

Pruski (2001) Micropropagation technology in early phases of commercial seed potato production. PhD Thesis, Wageningen University, Wageningen, p 166

Purseglove W (1972) Monocotyledons (tropical crops S). London: Longman Group Ltd

Purseglove W (1975) Tropical crop monocotyledonous. Longman Group, Britain
Ramanatha V, Matthews J, Eyzaguirre B, Hunter D (2010) The global diversity of taro: ethnobotany and conservation. Biodiversity International, Rome
Robert W, Cartwhight B (2006) Potato production. Oklahoma Cooperative Extension Service, Stillwater
Rudrappa U (2013) Yams nutrition facts and health benefits. http://www.nutrition-and you.com/yams.html
Rykaczewska K (2013) The impact of high temperature during growing season on potato cultivars with different response to environmental stresses. Am J Plant Sci 04:2386–2393. https://doi.org/10.4236/ajps.2013.412295
Sagoe R (2006) Climate change and root crop production in Ghana. Crops Research Institute, Kumasi
Sangakkara R, Attanayake B, Gajanayake N, Bandaranayake R (2004) Impact of mulches on growth and yields of cassava and sweet potato in tropical Asia. In: Proceedings of the 4th international crop science congress, Brisbane, 26 September–1 October 2004
Sarma A, Dutta T, Sarma A (1999) Effect of mulching technique with black plastic film (25 u) on potato crop under rainfed condition. Crop Res 18:383–386
Schüßler H, Gehrig H, Schwarzott D, Walker C (2001) Analysis of partial Glomales SSU rRNA gene sequences: implications for primer design and phylogeny. Mycol Res 105:5–15
Sefa-Dedeh S, Kofi-Agyir E (2002) Starch structure and some properties of cocoyam (*Xanthosoma sagittifolium* and *Colocacia esculentum*) starch and raphides. Food Chem 79:435–444
Shetto M (1999) Tillage practices and soil conservation measures in Tanzania. http://www.fao.org/ag/ages/3ero/Harare/partii/19Sheto.htm
Sieverding E (1991) Vesicular-arbuscular mycorrhiza management in tropical agrosystems. Deutche Gesellschaft für Technische Zusammenarbeit (GTZ), Eschborn. 371p
Sombroek WG, Gommes LR (1997) Changement de climat – Agriculture. In: FAO (ed) Changements du Climat et Production Agricole. FAO, Rome, p 472
Spooner DM, McLean K, Ramsay G (2005) A single domestication for potato based on multi-locus amplified fragment length polymorphism genotyping. Proc Natl Acad Sci 102:14694–14699
Sriom S, Mishra P, Rajbhar P, Singh D, Singh K, Mishra K (2017) Effect of different levels of nitrogen on growth and yield in potato (*Solanum Tuberosum* L.) Cv. Kufri Khyati. Int J Curr Microbiol App Sci 6:1456–1460. https://doi.org/10.20546/ijcmas.2017.608.428
Stokes C, Howden M (2010) Adapting agriculture to climate change: preparing australian agriculture, forestry, and fisheries for the future. CSIRO Publishing, Collingwood
Suyamto, Howeler H (2001) Cultural practices for soil erosion control in cassava-based cropping systems in Indonesia. In: Barker DH, Watson A, Sombatpanit S, Northcutt B, Maglinao AR (eds) In International Erosion Control Association, Ground and water Bioengineering for the Asia-Pacific Region. Science Publishers, Enfield, New Hampshire
Tailor J, Joshi H (2014) Harnessing plant growth promoting rhizobacteria beyond nature: a review. J Plant Nutr 37:1534–1571
Tshisola N (2014) Improved potato (*Solanum tuberosum* L.) seed production through aeroponics. MSc thesis, Stellenbosch University
Udo J, Ndon A, Asuquo E, Ndaeyo U (2005) Crop production techniques for the tropics. Concept Publications, Lagos
Ukom N, Ojimelukwe C, Okpara A (2009) Nutrient composition of selected sweet potato (*Ipomoea batatas* (L.) Lam) varieties as influenced by different levels of nitrogen fertilizer application. Pak J Nutr 8(11):1791–1795
Unger W (1994) Managing agricultural residues. Lewis Publishers, Boca Raton. 448pp
Uroz S, Calvaruso C, Turpault P, Frey-Klett P (2009) Mineral weathering by bacteria: ecology, actors and mechanisms. Trends Microbiol 17:378–387
Vernooy R, Sthapit B, Galluzzi G, Shrestha P (2014) The multiple functions and Services of Community Seedbanks. Resources 4(3):636–656
Vessey K (2003) Plant growth-promoting rhizobacteria as biofertilizers. Plant Soil 255:571–586
Wiebe K (2015) Strategic foresight: long-term projections from IFPRI, GFSF and AgMIP. Presented at the 12th Meeting of the Independent Science & Partnership Council held in Rome, 16 September 2015

Wiyo A, Kasomekerab M, Feyen J (1999) Effect of tied-ridging on soil water status of a maize crop under Malawi conditions. Agric Water Manag 45(2000):101–121

Woolfe A (1992) Sweet potato, an untapped food resource. Cambridge University Press, Cambridge, UK

8

Data Collection Using Wireless Sensor Networks and Online Visualization for Kitui, Kenya

Josephine Mbandi and Michael Kisangari

Contents

Abstract

Kenya is a developing country with a population of 47,213,282 people this comprises of 56% low-income earners. Small businesses and crop production represent 23% of the income within the country, which is at risk as soils become less productive. Various factors have led to this, climate change and land overuse being leading causes. Without adaptation, the rural to urban migration will continue to increase.

J. Mbandi (✉)
School of Computational and Communication Sciences and Engineering (CoCSE), Nelson Mandela Institution of Science and Technology (NM-AIST), Tengeru, Arusha, Tanzania
e-mail: mbandij@nm-aist.ac.tz

M. Kisangari
Centre of Excellence in Information Technology (CENIT@EA), Arusha, Tanzania
e-mail: Michael.kisangiri@nm-aist.ac.tz

Through Internet of Things (IoT) and specifically wireless sensor networks, we can change how we obtain and consume information. Small-scale farmers can collect data and in exchange receive useful information about their soils, temperature, humidity, and moisture content hence make better choices during crop production. Connected end devices bring in data, which is currently sparse in relation to small-scale farming. IoT will enable analysis and informed decision-making including crop selection, support equipment, fertilizers, irrigation, and harvesting. The cloud-based analysis will provide information useful for policy making and improvement.

This chapter presents a wireless sensor network (WSN) in mesh topography using XBee communication module, communication, and raspberry pi, combined with a cloud-based data storage and analysis. We successfully set up a proof of concept to test a sensor node that sends information to a RPi and onto an online visualization platform.

Keywords

IoT (Internet of Things) · WSN (wireless sensor networks) · Cloud computing · Climate change adaptation

Introduction: Background on Kitui County, Kenya

Kitui County is one of Kenya's 47 counties located in the lower eastern region of the country. The county has eight subcounties, namely, Kitui Central, Kitui South, Kitui East, Kitui Rural, Kitui West, Mwingi North, Mwingi West, and Mwingi Central. There are three livelihood zones in the county, marginal mixed farming, mixed farming livelihood zone, and formal employment, contributing 44%, 52%, and 4% of the population, respectively. The county has a population of 1,136,187 (Kenya National Bureau of Statistics (KNBS) 2019) people and covers an area of 30,570 square kilometers, of which 6,370 square kilometers is covered by Tsavo National Park (Magee et al. 2017). A large part of this County constitutes arid and semiarid land, and therefore over 50% of the population practice mixed farming and hence cultivate small pieces of land, and therefore the farming is heavily dependent on rainfall.

Climate Change and its Impact in the Region

Africa is considered to be one continent with the highest growth in population at 3% annually (IMF 2019). It is projected that by the year 2050, it will constitute about 25% of the world population (World Bank 2019). In addition, the mortality rate has reduced thereby straining available resources. Sustaining this growth has led to increase in social problems like rural to urban migration and lower yield in the food basket among others. As a result the majority of the active working force has moved to cities and towns to look for alternative sources of income.

The smallholder farmers have not been influenced by the "Green Revolution" as they have continued with the practice of nonuse of chemical fertilizers, pesticides, and genetically modified organisms. These inherent practices are referred to as "traditional farming." (Onwonga and Freyer 2006). In more recent years, farmers have become more reliant on extension officers. With the devolution of government services, extension officers have become few in some areas and highly concentrated in others (Franzel et al. 2014). These officers are expected to visit farmers to provide advisory services. It is usually tedious and time-consuming for both the farmer and the officer. There is thus need to investigate and keep record of weather conditions for better decision-making for small-scale famers who are the backbone of the economy in Kitui County (Magee et al. 2017). In addition assessment of cumulative health risk is a challenge in real-world situations because of limited data available for place and population-related pollution exposure (Solomon et al. 2016). With continued primary data collection, it is possible to create models that are more accurate in relation to this region.

The current weather stations within the country are a lot more concentrated in one half of the country and not so much on the other half. See map below (Fig. 1).

This study proposes to use a raspberry pi (RPi), a microcomputer quite popular now with IoT devices; it is of low-cost and easily adaptable to various conditions. The XBee is able to transmit information within short-range communication with very good accuracy and also low-power utilization and low-costs. The other parts of the paper will discuss methodology for the data collection, data transmission, storage, and visualization.

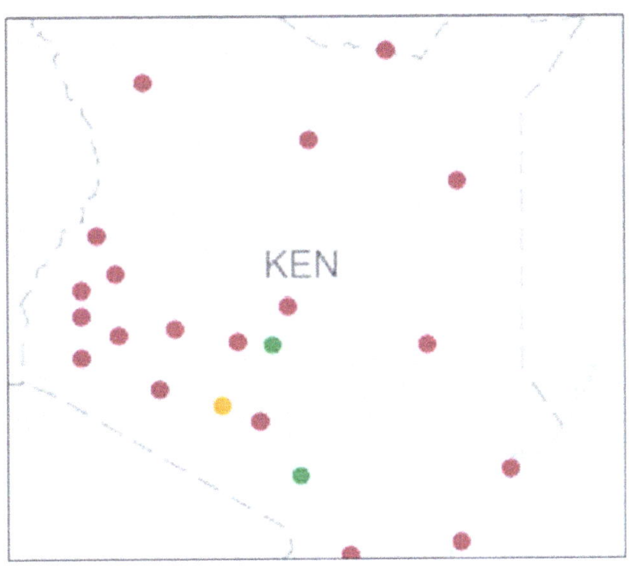

Fig. 1 Weather stations in Kenya (Solomon et al. 2016)

Other Studies

Studies have been done in agriculture and in pollution; however there is dearth in a studies combining pollution monitoring and agricultural parameters data collection. A review of previous studies has shown a gap in combination of these parameters which are of use to small-scale farming (Rehman et al. 2011) developed a WSAN with sensor and actuator nodes to provide controlled water supply; this is very geared toward precision agriculture. In 2016 (Raja et al. 2016) used various types of sensors to monitor temperature and moisture values of the soil. The data was shared to a central station periodically and concentrated on soil data. A similar study was undertaken in 2015 by (C and S 2015). This study builds on work done by (Rehman et al. 2011) where remote monitoring was done for data collection and information shared with research facilities (Vatsal and Bhavin 2017). Worked on a system showing real-time temperature and relative humidity and once again only concentrated in agricultural information for display locally.

There are several systems designed for precision agriculture. They, however, mostly solve the problem of irrigation or temperature control. Those in environment solve data collection challenges, while those in agriculture solve a single or at most two challenges.

Methodology

The Internet of Things (IoT) is described as connecting everyday objects via sensors and actuators to the Internet where the devices are intelligently linked together enabling new forms of communication between things and people and between things themselves. This technology has advanced rapidly in the last few years and has become quite popular in environment monitoring. The basic concept is illustrated in the figure below (Fig. 2).

IoT offers new and innovative ways to monitor and manage remote operations. "It allows having eyes and ears in remote places, constantly feeding applications and data stores with information. The low-cost of 'things' allows observing and

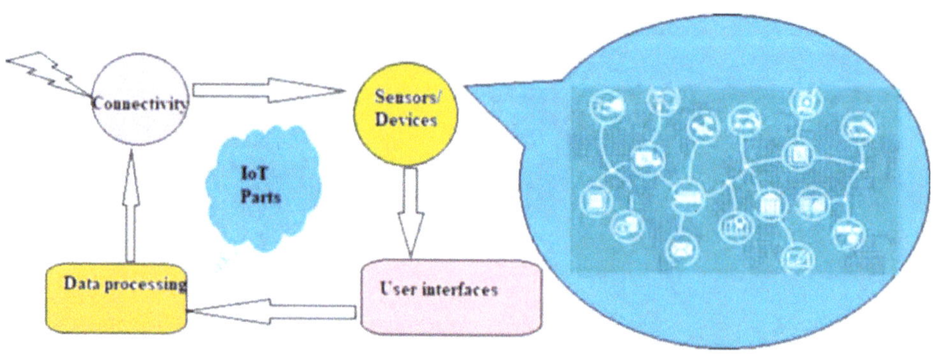

Fig. 2 Basic parts of an IoT system

managing activities that were previously out of reach. With the Internet of Things, it is also possible to collect information about events that were once invisible, such as correlating weather patterns with industrial production" ("IoT in action – Towards Data Science" 2020). The advancement in wireless sensor networks is used in monitoring and controlling of various aspect in an agricultural field. In the same light, weather aspects for a station field. A wireless sensor network is composed of hardware and software. These are low-cost and low-power smart sensor nodes which monitor physical conditions. In our initial prototype test case, we monitored temperature, moisture, and pollution. These nodes can be spread across a large geographic area (Fig. 3).

A node: The main components of a sensor node are a microcontroller, transceiver, external memory, power source, and one or more sensors. The controller performs tasks, processes data, and controls the functionality of other components in the sensor node. The sensor node or mote can be deployed infield to form wireless sensor network for crop management and environment monitoring system (Fig. 4).

This study uses ZigBee, an IEEE 802.15.4 for the wireless sensor network of nodes. It is common in networking standard for personal area networks. The physical layer of ZigBee operates in the unlicensed industrial, scientific, and medical radio bands of 868 MHz, 915 MHz, and 2.4 GHz depending on the region. In this study, we adopt 2.4 GHz band because it is unlicensed in Kenya. The ZigBee protocol

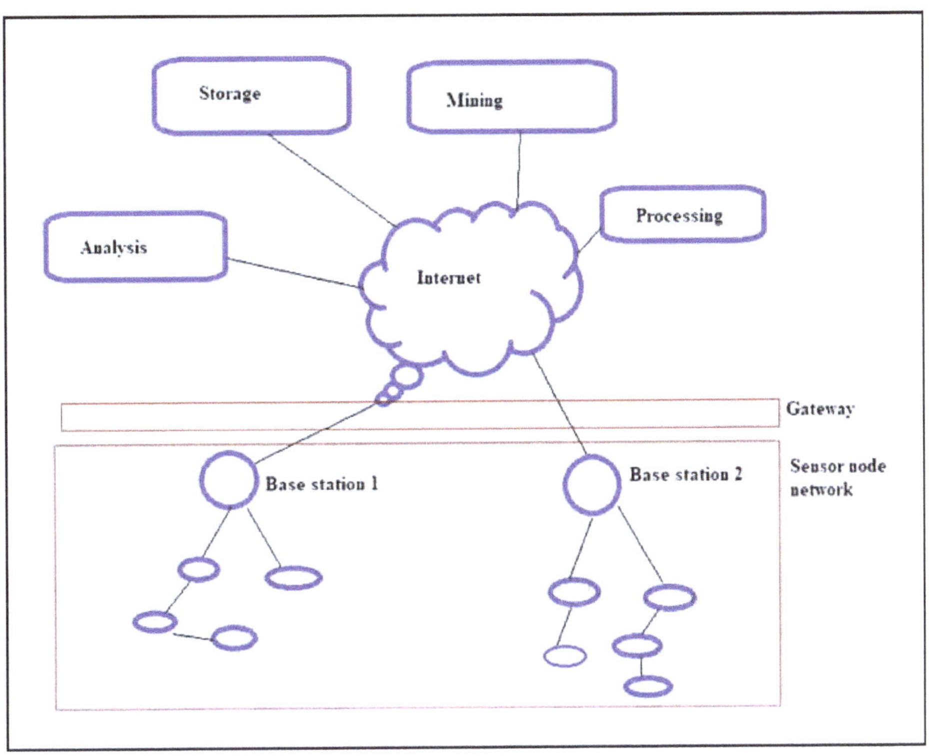

Fig. 3 Node based IoT system with offsite processing

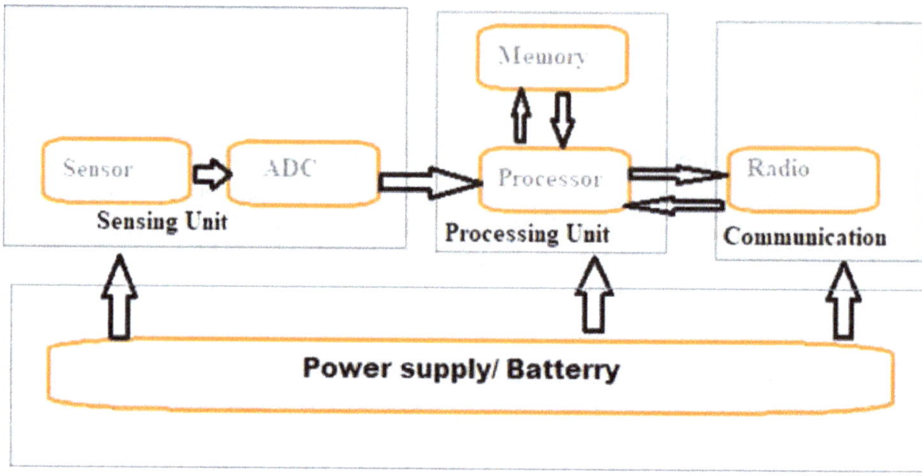

Fig. 4 Block diagram of a node

mainly focuses on low-cost and low-power consumption. The low-power consumption characteristic is necessary since sensors are usually placed at a remote location where battery power supply is virtually the only option. In order to attain a low-power consumption characteristic, the ZigBee protocol operates at low data rates (250 kbps at 2.4 GHz); this imposes its limitation where high data transmission applications are required. From node to the servers, it is thus more realistic to use other applications, for instance, Wi-Fi (802.11) which offer high data rates. This is only done from the coordinator node. Depending on the situation and environment, the networks can take three forms of topologies: star, cluster tree, and mesh. Our study uses the mesh topology. A mesh network is accomplished by allowing devices in the cluster tree to topology communicate with each other using multiple routes. Consequently, the devices are able to send and receive messages reliably even when their preferred path is down or congested. This is the major advantage of a mesh network over star and cluster-tree networks. However, a mesh network has no guarantee of bandwidth since no synchronization is used which requires disabling of beacon mode (Fig. 5).

Mesh topology, also referred to as a peer-to-peer network, consists of one coordinator, several routers, and end devices. The following are the characteristics of a mesh topology: A mesh topology is a multi-hop network; packets pass through multiple hops to reach their destination. The range of a network can be increased by adding more devices to the network. It can eliminate dead zones. A mesh topology is self-healing, meaning during transmission, if a path fails, the node will find an alternate path to the destination. Devices can be close to each other so that they use less power. Adding or removing a device is easy. Any source device can communicate with any destination device in the network. Compared with star topology, mesh topology requires greater overhead. Mesh routing uses a more complex routing protocol than a star topology" ("2.3 ZigBee Topologies | Introduction to the ZigBee Wireless Sensor and Control Network | InformIT" n.d.).

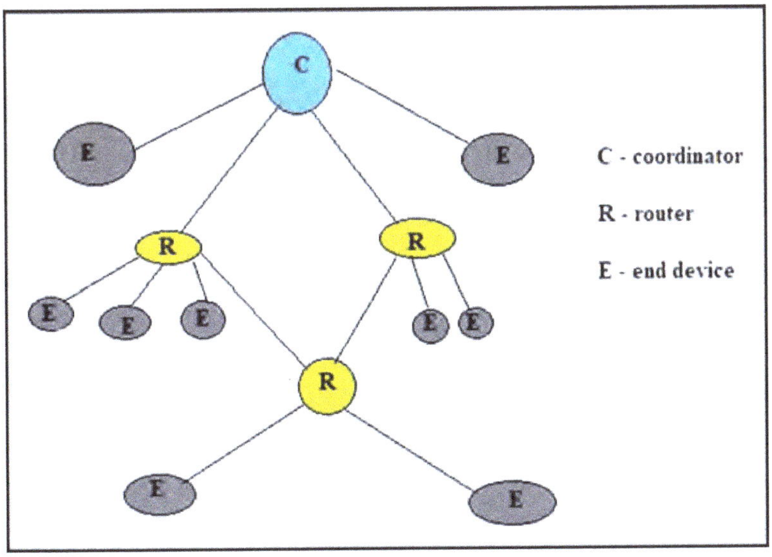

Fig. 5 Mesh-based network system

System Design

This study proposes the combination of solving challenges in environment and in agriculture with a single system at the same time contributing to data enhancement for policy issues in both precision agriculture and pollution primary data.

The building of pollution stations within the country present a problem of security and management. Having these stations placed at small farms and distributed keeps them safe. On the other hand, implementation of monitoring stations for local farmers is expensive and the communities lack the knowhow of management of the information collected. The cloud-based data management platform will provide this, while interested stakeholders (policy makers) will support the management of data. Kitui County is vast and most small farms lack connectivity to the main power grid. For the system to be self-sustaining in terms of power, the study uses solar photovoltaic and rechargeable batteries to power all electrical devices. The system incorporates a remote monitoring mechanism through a General Packet Radio Service modem to report soil temperature, soil moisture, WSN link performance, and photovoltaic power levels.

Having a system that encompasses data collection and data dissemination provides end to end solution for both small-scale farmers and policy makers.

The general workflow of the monitoring system consists of the following:

1. Taking temperature, humidity, and pollution data (PM 2.5 and PM 10) samples at predefined time intervals (Sensors)
2. Reading this data at a local level (Arduino Uno)
3. Sending data via a coordinator node (Raspberry Pi)
4. Visualizing data online

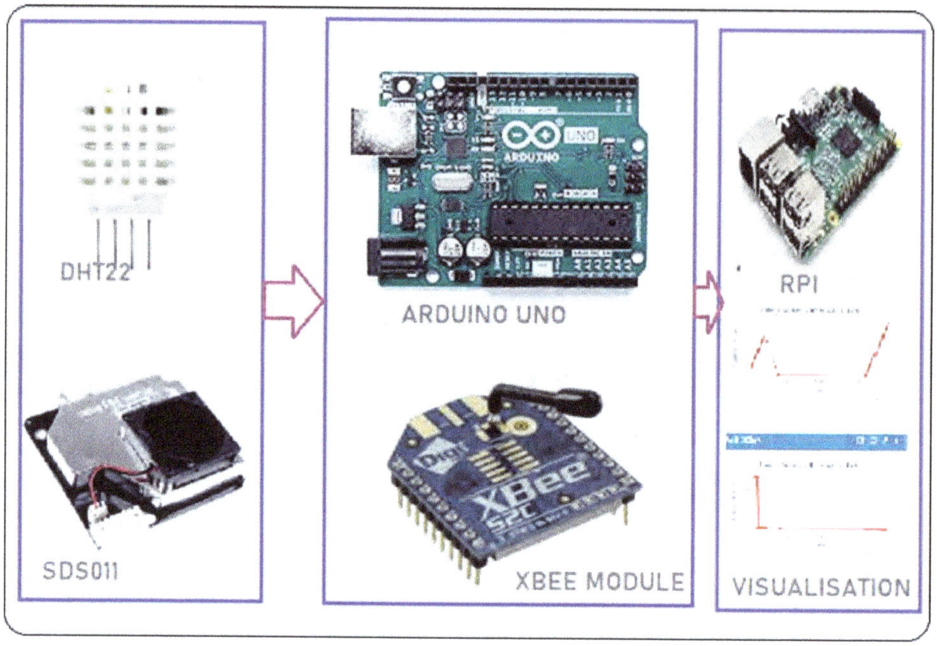

Fig. 6 Detailed block diagram of our system

5. Going to sleep
6. Waking up and repeating the previous steps (Fig. 6)

Hardware needed for the end point of the system:
For the sensor node:

1. Raspberry pi 3B – this is a low-cost, credit-card sized computer that plugs into a computer monitor or TV and uses a standard keyboard and mouse. It is a capable little device that enables people of all ages to explore computing and to learn how to program in languages like Scratch and Python (Raspberrypi.org n.d.)
2. Arduino UNO – Arduino is an open-source electronics platform based on easy-to-use hardware and software (Arduino.cc n.d.)
3. Humidity and temperature sensor – DHT22 Sensor
4. Pollution sensor SDS011
5. Connectors – female-female, male-female, and male-male
6. Resistors

Software used in this project:

1. XTCU – This was used in this system to setup the XBee radios so that they can transmit data between each other.
2. NOOBS – This is the operating system needed to setup a RPi.
3. Python.
4. JSON.

5. Arduino IDE.
6. ThingSpeak is a free online platform that was used for data visualization

System Implementation and Experiments

To set up and test this system, the XBee devices were first activated and set up using XCTU (a tool available online for free). They were all coded under the same network for ease of communication. The RPi was loaded with the necessary operating system (NOOBS) which comes with python – a coding language. In the project we also set up the Internet access for the raspberry pi as it acts as the gateway in this system.

On a laptop, we installed Arduino IDE (which is a development platform for Arduino boards) and also the sketches to run the two sensors for our prototype. JSON here is used as the shell to transfer data in a compact mode from the Arduino to the RPi.

On the RPi we run a program to call the data from the Arduino to the RPi and tested the same locally. Finally on ThingSpeak, we added the required APIs and run the visual data from our system.

The diagrams below further define our various steps (Figs. 7 and 8):

Once the codes were tested and found to be producing steady outputs, we put the sensors in a monitored environment – domestic fridge with proper calibration to test the humidity and temperature parameters and the results were displayed on the local Arduino software as 169% humidity and 12° Celsius continuously and repeatedly (Fig. 9).

```
#include<SoftwareSerial.h>
#include <SDS011.h>
#define Tx, Rx;
float p10,p25;
int error;

SDS011 my_sds;

void setup() {
  my_sds.begin(Tx,Rx);
  Serial.begin(9600);
}

void loop() {
  error = my_sds.read(&p25,&p10);
  if (! error) {
    Serial.println("P2.5: "+String(p25));
```

Fig. 7 Figure 15 – clips of Initial codes for SDS sensor

```
#include <dht.h>

#define dht_apin A2 // Analog Pin sensor is connected to

dht DHT;
void setup(){

  Serial.begin(9600);
  delay(500);//Delay to let system boot
  Serial.println("DHT11 Humidity & temperature Sensor\n\n");
  delay(1000);//Wait before accessing Sensor

}//end "setup()"

void loop(){
  //Start of Program
     DHT.read11(dht_apin);
```

Fig. 8 Clips of Initial codes for DHT sensor

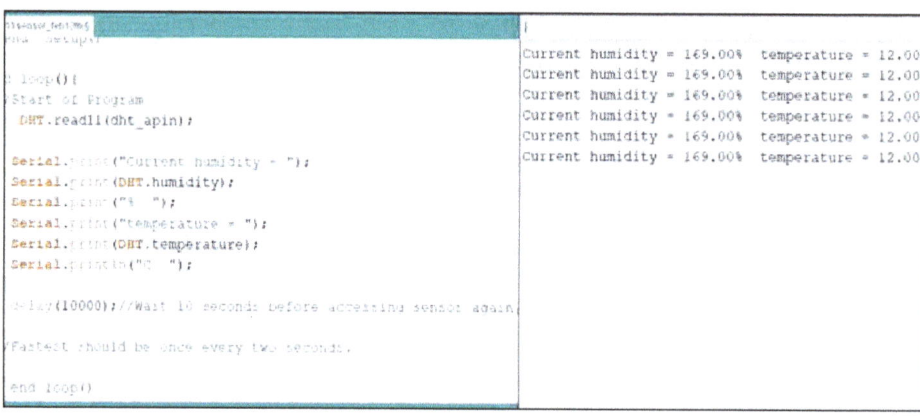

Fig. 9 Testing the DHT11 for both humidity and temperature in a controlled environment

We then transferred the tests to the natural environment and connected the router XBee connected to the raspberry pi. We were able to read the results from the temperature and humidity sensor as well.

We also run similar tests on the SDS011 – pollution sensor and obtained a consistent flow of results as below.

Once this was done, we then connected the end nodes to the router node for data aggregation. This was done using the python on the raspberry pi and wrapping the file using Json. We could then see our data collected using the IDs EN1, EN2, and EN3 from the three Arduino/XBee connected end nodes (Figs. 10, 11, 12, and 13).

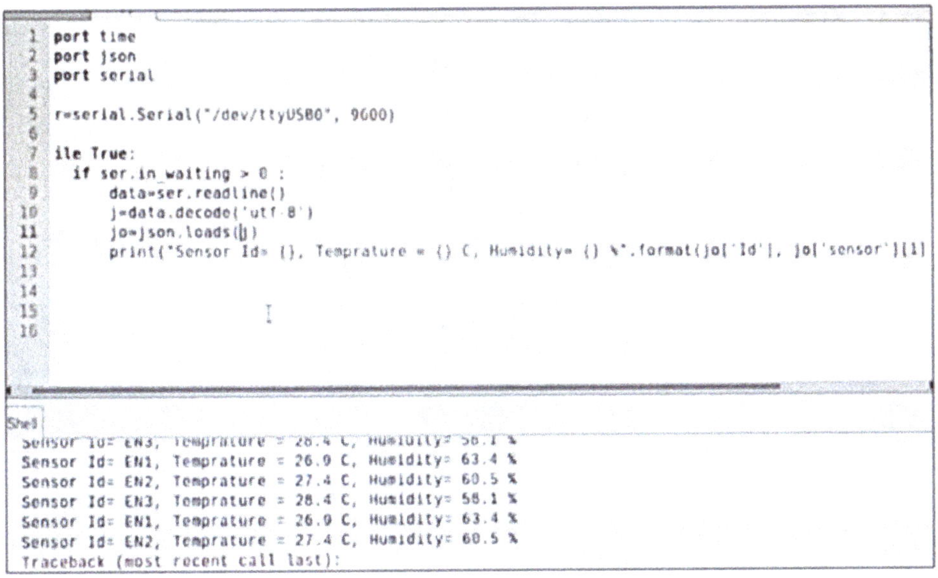

Fig. 10 Data accessed locally on RPi from three Arduino stations with DHT sensor only

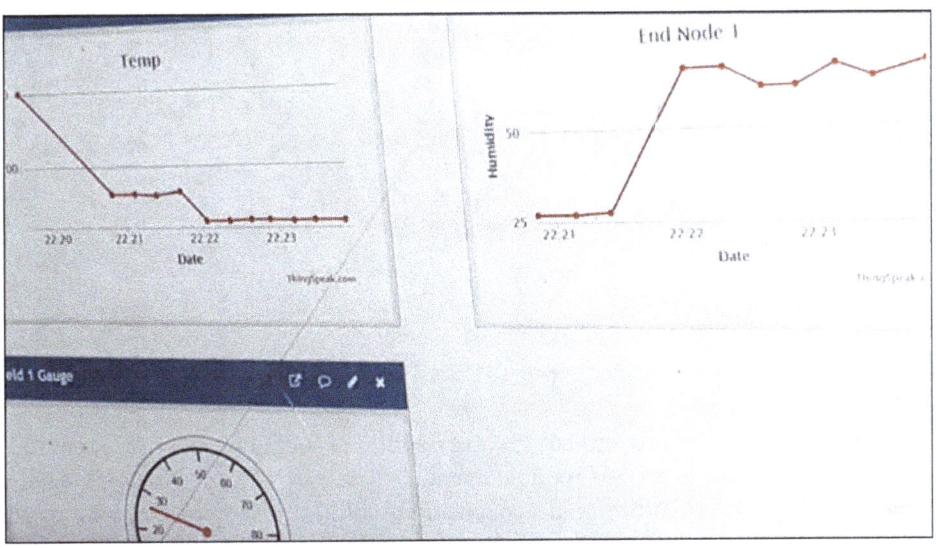

Fig. 11 Display on ThingSpeak of test data – screen shot taken during the testing of the prototype

Conclusion

The phase of the entire project included:

1. Coming up with a design to collect data from sensors via a sensor node.
2. Sending the data to an online visualization platform via a control node and this step has been successful.

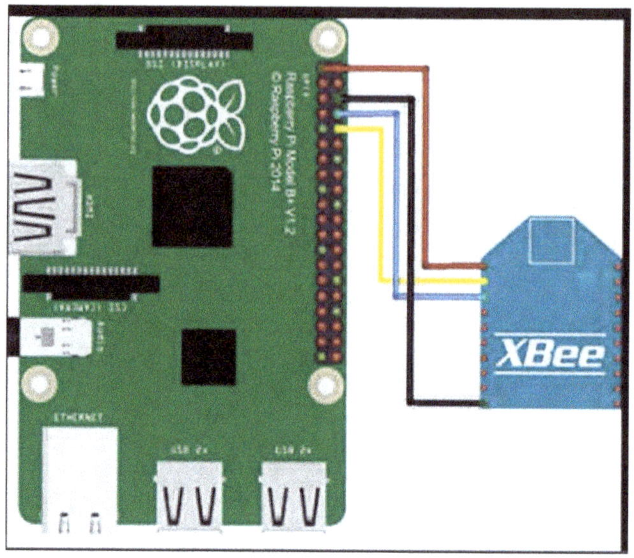

Fig. 12 XBee connection to RPi (drawn using Fritzing software)

Fig. 13 Sensor node using Arduino, XBee, DHT22, SDS011

Continuing work in this project involves adding a local database for localized data storage and adding a GSM module that allows ease of access of the online data remotely. The online platform and visualization will also be improved to be more robust with a private platform that can be more scalable in the future.

References

2.3 ZigBee Topologies | Introduction to the ZigBee Wireless Sensor and Control Network | InformIT (n.d.) Retrieved January 15, 2020, from http://www.informit.com/articles/article.aspx?p=1409785&seqNum=4

Arduino.cc (n.d.) Arduino – Introduction. Retrieved May 7, 2020, from https://www.arduino.cc/en/guide/introduction

C, A, & S, A (2015) A study on developing a smart environment in agricultural irrigation technique. Int J Ambient Syst Appl 3(2/3):11–17. https://doi.org/10.5121/ijasa.2015.3302

Franzel S, Sinja J, Simpson B (2014) Farmer-to-farmer extension in Kenya: the perspectives of organizations using the approach. ICRAF. https://doi.org/10.5716/WP14380.PDF

IMF (2019) World economic outlook (October 2019). Retrieved January 21, 2020, from https://www.imf.org/external/datamapper/datasets/WEO

IoT in action – Towards Data Science (2020) Retrieved January 15, 2020, from https://towardsdatascience.com/iot-in-action-a8b7fac83619

Kenya National Bureau of Statistics (KNBS) (2019) 2019 Kenya population and housing census volume I: population by county and sub-county (Vol. I). Retrieved from http://www.knbs.or.ke

Magee RV, Beerbower E, Bodhi B, Schindler S, Pfattheicher S, Reinhard M-A, … Gorski PC (2017) KITUI county 2016 short rains food security assessment report. ABA J 102(4):24–25. https://doi.org/10.1002/ejsp.2570

Onwonga R, Freyer B (2006) Impact of traditional farming practices on nutrient balances in smallholder farming systems of Nakuru District, Kenya. In Proceedings of the tropentag conference October, (Octuber 2014), 11–13. Retrieved from https://www.researchgate.net/publication/228372444_Impact_of_Traditional_Farming_Practices_on_Nutrient_Balances_in_Smallholder_Farming_Systems_of_Nakuru_District_Kenya/link/00b495156e425eb5ba000000/download on January 19, 2020

Raja R, Aruna P, Rajeshwaran N, Vignesh M, Vijayakumar M, Rajeshwaran N, … Coimbatore T (2016) Automatic irrigation control using wireless sensor network 1299–1302

Raspberrypi.org (n.d.) What is a raspberry pi? Retrieved May 7, 2020, from https://www.raspberrypi.org/help/what-is-a-raspberry-pi/

Rehman A, Ali G, Shaikh NA, Shaikh ZA (2011) Efficient irrigation water management using cost effective wireless sensor and actuator networks. Sarhad J Agric 27(3):423–435

Solomon G, Morello-Frosch R, Zeise L, Faust JB (2016) Cumulative environmental impacts: science and policy to protect communities. Annu Rev Public Health. https://doi.org/10.1146/annurev-publhealth-032315-021807

Vatsal S, Bhavin M (2017) Using raspberry pi to sense temprature and relative humidity. Int Res J Eng Technol (IRJET) 4(2):380–385. Retrieved from https://irjet.net/archives/V4/i2/IRJET-V4I276.pdf

World Bank (2019) World's population will continue to grow and will reach nearly 10 billion by 2050. Retrieved January 21, 2020, from http://blogs.worldbank.org/opendata/worlds-population-will-continue-grow-and-will-reach-nearly-10-billion-2050

9

Climate Change Impact and Adaptation: Lagoonal Fishing Communities in West Africa

K. Sian Davies-Vollum, Debadayita Raha and Daniel Koomson

Contents

K. S. Davies-Vollum (✉) · D. Raha · D. Koomson
Environmental Sustainability Research Centre, University of Derby, Derby, UK
e-mail: s.davies-vollum@derby.ac.uk; d.raha@derby.ac.uk; d.koomson@derby.ac.uk

Abstract

Lagoons are a common feature of the low-lying West African coastline. These lagoons are resource-rich and biodiverse. The small-scale fishing communities, which border them, are dependent on the resources and ecosystem services for their livelihoods and well-being. Climate change has had significant and diverse effects on both the lagoons and their surrounding communities. Sea level rise has caused erosion of the coast and increased the risk of floods. Changes to rainfall patterns have caused shifts in lagoon ecosystems and physical cycles. Of particular relevance to lagoon fishing communities is the fluctuation in quantity and distribution of fish catch that they rely upon for economic livelihood. Understanding the vulnerability of these communities to the effects of climate change is critical to supporting and developing successful adaptations. Using a case study from Ghana, sustainable livelihoods approach (SLA) and vulnerability framework are used to characterize the community vulnerability, giving insight into the temporal and spatial dynamics of vulnerability and how subsections of the community may be identified and prioritized for adaptation interventions. A scalar analysis of the relevant coastal and environmental frameworks and policy to support climate change adaptation in coastal communities reveals the common challenges in implementing adaptation interventions and strategies in the region. A policy gap exists between high level, institutional coastal, and climate directives and implementation of climate adaptations at the local level. That gap might be bridged by a participatory approach that places coastal communities at the center of creating and enacting climate change adaptations.

Keywords

Lagoon · Vulnerability · Adaptation · Fishing communities

Introduction

Coastal lagoons, and the wetlands that are often associated with them, are estimated to occupy 13% of coastlines worldwide (Duck and De Silva 2012). They are found at low-lying, sandy coastlines where they support highly productive ecosystems that provide a wide range of natural functions and ecosystem services to the communities living around them. Lagoons along African coastlines are often home to fishing communities, whose livelihoods are entirely dependent on the resources supplied by them. In West Africa, lagoon communities take advantage of highly productive coastal fisheries that are supported by upwelling along the Gulf of Guinea, which brings nutrients to surface waters. The fishing communities here tend to fish in both shallow coastal waters and coastal lagoons (also known as coastal lakes). Climate change is affecting lagoons and the fishing communities that live around them, in multiple interconnected ways. Sea level rise has eroded coastal land and increased the risk of flood. Changes to rainfall and temperature patterns have altered natural lagoon cycles and coastal ecosystems. The resultant changing ocean and lagoon conditions have altered fish catch and fishing practice. The effects of climate change thus have significant implications for the well-being and livelihood of lagoon fishing communities. Understanding how climate change affects lagoon fishing communities is critical to supporting and developing successful adaptations. In this chapter, the impacts of climate change on the lagoons of West Africa, specifically in Cote d'Ivoire, Ghana, Benin, Togo, and Nigeria (Fig. 1), and the fishing communities they support are discussed. How the vulnerability of these communities is characterized, assessed, and connected to climate change is explored. A case study of a fishing community in Ghana is used to exemplify how understanding vulnerability is critical for effective and nuanced community-level adaptation planning. Current international, regional, and national frameworks, governance, and policy to support the creation and implementation of successful adaptation strategies for lagoon communities are presented. The gap between coastal- and climate-related policy and successful implementation of adaptations to climate change in coastal communities is considered and the participatory approach required to bridge that gap presented.

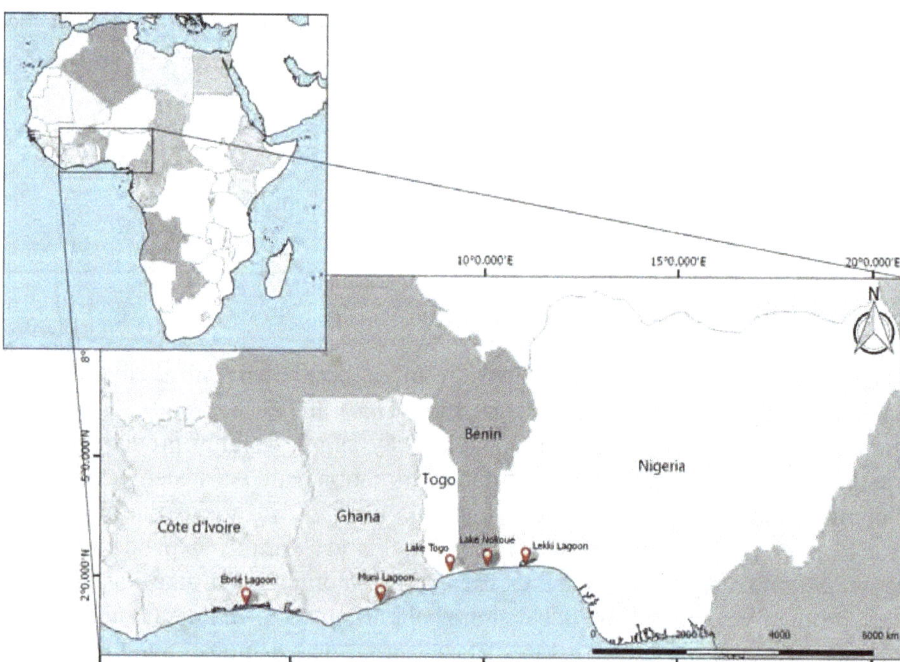

Fig. 1 Map of West African coastal countries with key urban area and lagoons shown

Lagoon Environments and Communities

Lagoon Environments

Lagoons are highly dynamic coastal environments, vulnerable to changes in oceanic and meteorological conditions associated with climate change. They are bodies of coastal water fully or partially separated from the open ocean by a natural barrier. This barrier is of low elevation, often less than two meters above sea level, and consists of sand that is easily eroded and transported. Lagoon barriers can entirely separate the lagoon from the ocean, or they can have one or more breaks, known as breaches or tidal inlets, which connect the lagoon to the ocean. The breaks are important as they act as channels that allow ocean water and biota to enter the lagoon. The barriers can change from whole to breached, and vice versa, in response to variations in weather conditions, hydrological conditions, availability of sediment, and sea level. A seasonal cycle of opening and closing of the sand barriers is common. Natural breaching usually results from overtopping of the barrier from the lagoon when the lagoon is full of freshwater at the end of rainy season. The opening of a lagoon to the ocean allows seawater to flow into the lagoon, resulting in rapid shifts in water temperature, salinity, dissolved oxygen content, and concentration of anthropogenic pollutants. Opening can also alter the tidal amplitude and influence in the lagoon as well as the amount and location of erosion and sediment deposition. Opening can also have an effect on the incidence, amount, and location of flooding. Such fluctuations in the physical environment can create unstable and stressful conditions for lagoon organisms at the species, community, and ecosystem level that can then negatively affect their health and productivity. This in turn influences the availability of natural resources, including fisheries.

Lagoons, or coastal lakes as they are sometimes known in the region, are found at low-lying coastal areas throughout Africa. Along the West African coastline of focus here, there are more than 200 lagoons from Cote d'Ivoire to western Nigeria. This coastline experiences the Guinea Ocean current, which flows from west to east along the west-east trending coastline, creating similar ocean conditions in the region. The lagoons along this coastline vary in size and shape as well as the nature of their connection to the ocean. Two of the largest are Lake Nokoué in Benin and the Lekki Lagoon complex in Nigeria (Fig. 1). Lake Nokoué lagoon covers an area of 120 km^2. A former estuary of the Ouémé River, a sand barrier has now enclosed the estuary (Anthony et al. 2002), preventing permanent connection to the sea. The Lekki-Lagos lagoon complex in Nigeria parallels the coastline for approximately 200 km (Bernacsek et al. 1990) incorporating two separate lagoons: the Lekki and Lagos lagoons. The majority of lagoons, however, in the region are much smaller than these two, with an areal extent from a few to tens of square kilometers.

Lagoons have high biodiversity in tropical and subtropical regions and are often the site of species-rich wetlands. West African lagoons are no exception with high productivity and biodiversity. The key natural resources of these biodiverse lagoons are fisheries and mangroves. According to the USAID, lagoon settings play a key role to fisheries in the area and have been recognized as contributing $400 million annually to the regional economy (USAID 2014). However, the West Africa Regional Fisheries Program (WARFP) indicates that incomes for artisanal (small-scale) fishers have dropped by as much as 40% over the last decade (The World Bank 2015). This has knock-on effects for the livelihood and well-being of these small-scale fishing (SSF) communities, with the potential to increase their vulnerability to the impacts of climate change. Mangroves are the key lagoon wetland vegetation and provide a variety of ecosystem services: shoreline protection and storm buffering through trapping sediment and dissipating wave energy, regulation of water quality, and provision fish habitats and nurseries (Blankespoor et al. 2014). Through the ecosystem services they provide, mangroves are estimated to contribute approximately 3 billion US dollars annually to the West African economies (Boateng 2018). Despite the importance of mangroves to lagoon communities, they have been decimated due to overexploitation, clearance, and pollution (Boateng 2018; Djihouessi et al. 2016). The impacts of climate change at the coast (sea level rise, changes to temperature, shifts in rainfall patterns, and increased storminess) have further contributed to their demise.

Lagoon Communities

Coastal towns are Africa's most developed, most densely populated urban areas (State of African Cities 2008), and migration to the coast has been and remains a long-term trend (Ochiewo 2004). The major cities of Abidjan (Cote d'Ivoire), Accra (Ghana), Cotonou (Benin), Lome (Togo), and Lagos (Nigeria) are all situated around one or more lagoons (Fig. 1). Cote d'Ivoire has a coastal population of 8.2 million, which accounts for 36% of its total population (World Bank 2019). The economic capital and largest urban area of Cote d'Ivoire is Abidjan, which is built around the Ébrié Lagoon. Seventy percent of Ghana's approximately 25 million population (Ghana Embassy 2014) lives in the southern part of the country with maximum population density at the coast. The Korle Lagoon has played a significant role in the development of its capital, Accra. Benin has a short coastline compared to the total area, yet half of its population live in Cotonou, the capital, which is located around Lake (lagoon) Nokou (World Bank 2019). In Togo, the City of Lome lies on the eastern shores of Lake (lagoon) Togo. Lagos, Nigeria's largest city, sits on the Western shores of Lagos Lagoon, part of the Lekki lagoon-barrier system. These lagoons have been an important source of natural resources, including fisheries, supporting the development of the surrounding urban areas. Fishing communities have traditionally lived around the shores and along the lagoon barriers of these highly urbanized areas. Despite the rapid growth and population increase of urban coastal areas, fishing communities still have a strong presence. For example, in Cotonou fishing remains a mainstay of the economy (Dossou and Glehouenou-Dossou 2007). Lagoons located in less urbanized (rural or peri-urban) areas of the coast are also home to large numbers of small fishing communities. Between 40% and 45% of the rural work force of Lagos State of Nigeria engage in SSF, and it is the dominant type of employment (Akanni 2010). At Aby Lagoon in the Cote d'Ivoire, fishing is noted as the most important activity for the economy and livelihood of families in the region (Njifonjou et al. 2006), and in Ghana 66.5% of household livelihood activities in coastal communities are classified as fisheries-based (Evadzi et al. 2018). No matter where they are located, whether in urban or rural areas, these fishing communities are dependent on lagoon and ocean fish stocks for their livelihood and food security. In addition, lagoon fishing communities depend on the myriad ecosystem services that a lagoon system provides such as water resources, flood mitigation, washing and bathing, and wood resources, all of which are subject to the impacts of a changing climate.

Impacts of Climate Change on Lagoons and Lagoon Communities

Sea Level Rise

Climate change negatively affects lagoons and the communities they support through sea level rise, changes to rainfall, and increased ocean and air temperatures. During the twentieth century, global average sea level rose by 3.2 mm/year (Rahmstorf 2007). During the twenty-first century, rates of SLR are expected to increase with estimates varying greatly from 0.4 meters to 2.5 meters (Hauer et al. 2019). Historic data for SLR in Africa has been limited, and its impacts have been understudied across the continent at both national and sub-national levels (Brown et al. 2011). However, there are an increasing number of studies relating to SLR and its impacts along the coast of West Africa. Coastal management in Ghana has focused on physical processes; thus, most information on SLR and erosion in the region is from here. The average rate of sea level rise over the twentieth century measured at the Takoradi tidal station in western Ghana is 3.34 mm/year (Sagoe-Addy and Appeaning Addo 2013), and by inference, the rate along this coastline is likely to be similar. Evadzi et al. (2018) noted that on average, sea level in Ghana has risen by about 5.3 cm over the last 21 years, and SLR accounted for approximately 31% of the observed annual coastal erosion rate (approximately 2 m/year) in Ghana. The lagoon-barrier coastlines of West Africa are particularly susceptible to erosion associated with SLR because of their low coastal topography and their composition of loose, easily erodible sand. Sea level rise can cause erosion of lagoon barriers, increasing the probability of lagoon opening to the ocean. Average rates of coastal erosion for the coastline between Cote d'Ivoire and Nigeria are around 1.5 meters/ year (World Bank 2019; Wiafe 2010). However, rates of erosion vary greatly along the coast with erosion hotspots recording more rapid shoreline loss. In eastern Ghana, close to the border with Togo, rates of erosion of up to 3.9 m/year have been recorded (Boateng 2012), and up to 4 meters/year has been recorded along parts of Benin's coastline (World Bank 2019). High rates of erosion have also been observed near the Lekki lagoon, along Nigeria's lagoon-barrier coast (Danladi et al. 2017). At all these locations, rates of erosion are often exacerbated by the loss of mangroves (Diop and Sherin 2016; Boateng 2018), and in a negative feedback loop, coastlines with high erosion rates often experience increased mangrove loss as the root structures of the plant become compromised due to the loss of substrate through erosion. Erosion associated with SLR affects not only natural coastlines, including the lagoon barrier integrity, but also the stability of human structures. This includes dwellings but also fishing-related structures. In Ghana, the loss of fish landing sites in Accra is already recognized (Appeaning-Addo 2014) resulting in the collapse of artisanal fishing.

Flood Risk

Flood risk for lagoon communities is enhanced by sea level rise, and there are multiple examples of this recorded for the countries considered here. Several Nigerian coastal states are recognized as having high flood risk for an SLR of less than a meter with the barrier lagoon coast of Lagos-Badagry-Seme in the southwestern part of Nigeria noted as particularly vulnerable (Odunga et al. 2014). Also in Nigeria, flooding and associated intrusion of salt water from the ocean have been recorded at the Lekki lagoon-barrier system since the early 1990s (Awoskia and Ibe 1993). The lagoon at Abidjan was highlighted as having a high risk of flooding associated with SLR (Adelekan 2010). In Ghana's central coast, a study of the Muni Lagoon indicated that a 1 meter rise in sea level would almost completely submerge its barrier and inundate the lagoon (Davies-Vollum and West 2015). The poorest communities are those most at risk from flooding, which includes those engaged in SSF. Temporary relocation is the main strategy adopted by members of the communities to adapt to coastal flooding. Some communities also use cheap materials for the construction of their houses as an adaptation to minimize economic loss during inundation (Evadzi et al. 2018). Increased water salinity associated with SLR and inundation inhibits biota in coastal ecosystems, including mangroves, and in a study of coastal wetlands in LEDCs, Blankespoor et al. (2014) estimated that up to 68% of these would be lost if subject to a 1 m rise in sea level.

Rainfall and Temperature

Rainfall projections for the West African coast indicate that during the twenty-first century, precipitation will become extreme with dry seasons becoming drier and wet seasons wetter. Along the West African coast, the major rainy season occurs from approximately March to July, with some geographic variation. Precipitation changes for West Africa include a projected variability in seasonal monsoonal rains (Adeniyi 2016), and there is evidence that the timing and amount of rain that falls during the rainy season have changed (Dossou and Glehouenou-Dossou 2007). In southwestern Nigeria, the rainy season appears to have shifted from March–November to May–October (Adelekan and Fregene 2015), and in Ghana, there seems to be less predictability in the onset of monsoonal rains (Koomson et al. 2020). Changes to the amount and timing of precipitation during the rainy seasons can modify the amount of freshwater input to lagoons and disturb the natural cycle of lagoon opening. Increased rainfall during the wet season results in more water impounded behind the barrier. This not only causes flooding but also increases the probability of lagoon waters overtopping and/or breaking through the barrier. Rainfall that is more evenly distributed throughout the year results in decreased risk of overtopping and lagoon opening. More evenly distributed rainfall patterns also lead to a more constant input of freshwater, increasing homogeneity of lagoon water conditions throughout the year and stabilizing salinity fluctuations. This in turn affects lagoon ecosystems including fish stocks (Diop and Sherin 2016).

Lagoon and shallow marine ecosystems are also affected by changing water temperatures. A general warming of coastal waters has been noted for the Gulf of Guinea (Druyan 2011). Although there is limited data on the impact of this warming on fisheries in the region, marine species in equatorial regions in general have seen a decline, as they are unable to respond to rapidly increased water temperatures (Hastings et al. 2020). Deoxygenation and acidification of ocean waters are also a consequence of climate change, and both are expected to indirectly reduce maximum fish catch. How fishing communities deal with the physical and ecological changes related to climate change is dependent on their vulnerability to those changes and their ability to adapt to them. Vulnerability and adaptation are inextricably linked, and any discussion about climate change adaptations warrants an overview of the architecture of the vulnerability of lagoon fishing communities.

Vulnerability of and Adaptation to Climate Change in Lagoon Fishing Communities

Vulnerability can be defined in several different ways in line with different epistemological orientations. However, whether grounded in physical science and human or political ecology, the key concept in vulnerability is the potential for loss or susceptibility to harm. The vulnerability of a community to climate change can be considered as its susceptibility to all the collective changes it experiences. The vulnerability of lagoon fishing communities is related to several multi-scalar factors or causes. The causes of vulnerability in lagoon fishing communities might be categorized into four sources: international, intrinsic, internal, and institutional. International sources refer to global climate and physical environmental change processes such as sea level rise, increase in sea surface temperature, and others, as outlined in the preceding sections. These processes, inter alia, reduce fish catches and revenue (Lam et al. 2016; Dey et al. 2016). Reduced incomes, in neoclassical economic terms, imply reduced purchasing power and so low material well-being (i. e., possession of natural, physical, financial, human resources). Intrinsic sources of vulnerability are those unique to coastal fishing communities and directly linked to their fishing activities. These factors are diverse and include the hazards encountered during fishing (e.g., storms), the seasonality of fish catches, as well as the high unpredictability of making catches. It also includes their high susceptibility to macroeconomic changes. For example, they are forced to accept the market cost of fuel and other necessary resources.

While intrinsic sources refer to features, which are uniquely associated with the small-scale fishing livelihood itself, internal sources refer to features commonly associated with coastal fishing communities. These include overfishing, gender and income inequality, intra-community conflicts, lack of cooperation, mistrust, and limited livelihood diversification opportunities (Béné 2006). Because artisanal fishing is mostly done in crews and lagoon fishing communities are usually small, trust, camaraderie, and solidarity are essential to sustaining the fishing activity and the community. When these sociocultural factors are missing, the very survival of the fishing livelihood is threatened. Moreover, these internal factors create differentiated experiences of vulnerability within fishing communities (Zacarias 2019; Koomson et al. 2020). Lastly, institutional sources are those related to rural planning, infrastructural development, and encroachment of industrial trawlers. Fishing communities do not usually participate in fisheries planning and management decision-making, have little social amenities, and often must deal with encroachment of large trawlers in local fishing waters. This list of vulnerability sources is by no means exhaustive. It is only a sample of the common features of artisanal lagoon fishing communities, which interact in complex ways with the impacts of climate change to make them susceptible to socioeconomic losses.

Whereas vulnerability refers to the susceptibility to harm, adaptation is the process of adjusting to the actual or expected harm and associated impacts, in order to either moderate the harm or exploit beneficial opportunities. The term adaptive capacity is used when considering how capable an individual or community is of adapting to harm or change. The adaptive capacity is based on an individual or a community's available resources that can be drawn on to achieve adaptation. The specific ways in which those combinations of available resources are enacted to moderate harm or exploit beneficial opportunities are then referred to as adaptive strategies. When considering and assessing adaptations and adaptive capacity, there are several critical concepts that should be accounted for. Firstly, it is rooted in the definition of adaptation. An adaptation is an adjustment to moderate actual or potential impacts, not to eliminate or "overcome" them. Secondly, the definition itself makes no value judgment about the legality of the adjustment process or on how the moderation of actual or potential impacts is achieved. In normative terms therefore, any action that has some moderating effect on impacts would qualify as an adaptation strategy. Thirdly, from the definition, adjustments to moderate impacts or benefits from opportunities would not be limited to actions only but also prohibitions. Communities themselves may spontaneously enact adaptations and adaptive strategies. These are referred to as autonomous adaptations.

Community-Level Autonomous Adaptations

Autonomous adaptations enacted to reduce or moderate harm in SSF communities might also be considered in the context of reducing risk to those communities. Risk reduction strategies that are spontaneously developed by communities whose livelihoods depend on natural resources, such as SSF communities, can be divided into five categories: over time (*storage*), across space and time (*mobility*), across asset classes (*diversification*), across households (*communal pooling*), and *market exchange* (Agrawal 2008). Communities and individual households may choose to combine and enact any of these strategies to achieve their desired outcomes. This is central to the concept of sustainable rural livelihoods and is important to examine when considering adaptations and adaptive capacity.

One of the earliest discussions on sustainable livelihoods (Chambers and Conway 1992) provided a similar list, which also included reduction of current consumption (*stinting*), preservation and protection of the asset base for recovery (*protection*), and making claims on relatives or NGOs (*claiming*). In a similar discussion on sustainable livelihoods, Scoones (1998) describes two more livelihood strategies: increasing input per unit area (*intensification*) and expanding the area under production (*extensification*). These livelihood/adaptation strategy typologies were developed predominately from the study of rural agrarian communities. However, they are also useful for characterizing adaptation strategies reported for commonly reported stressors experienced by fishing communities worldwide, not just in West Africa. These stressors are noted in Table 1. Not all stressors listed are directly related to climate change. Physical stressors like storms and floods and decreasing fish catches are a consequence of climate change. However, encroachment of industrial trawlers into artisanal fishing zone causing competition is not. Competition from trawlers is noted here because of the large impact it has on SSF communities and their vulnerability to climate change. An "experimentation" grouping in the table captures adaptation practices that are less well-established and in an initial learning phase of development (Table 1).

The most common adaptation strategies adopted by SSF communities are intensification strategies (Béné 2006). When fish catches are low, the first and immediate response of SSF communities is to put in more effort either by going further out at sea, spending longer hours at sea, changing fishing gear (equipment) to target fish species that may be relatively more abundant, or increasing the frequency of fishing activities. For example, in a coastal lagoon fishing community situated on the Muni Lagoon in Ghana, the major fishing season, which traditionally lasted from August to April, has been extended by some fishing crews by starting their fishing activities from July to May of the following year. This is largely in response to an erratic rainfall pattern in the major rainy season in southern Ghana. For supplemental incomes, lagoon communities may increase fishing in lagoons and use gears or methods considered illegal. In Benin, Acadjas (a purely ingenious practice of creating a favorable microclimate within a lagoon with sticks and leaves for recruitment of big fish) and médokpokonou (another traditional method which uses nets with fine and tight stitches) are both prohibited, but they are commonly practiced. This is because the former reduces overall fish species richness, and the latter removes small and juvenile fishes (Dossou and Glehouenou-Dossou 2007; Niyonkuru and Lalèyè 2010). The use of mosquito nets in fishing communities has also been observed in Cote d'Ivoire, Ghana, and several other sub-Saharan African countries (Short et al. 2018). This diverts the use of nets away from their role in protecting coastal communities from mosquitoes and negates efforts to reduce

Table 1 Examples of autonomous adaptation to key stressors of small-scale fishing livelihood in West Africa

Stressor	Examples of adaptations
*Physical (*e.g., *storms, floods and SLR)*[a]	Use of sandbags and rubbish to form defense walls Change of building materials (from wood and thatch to cement blocks and galvanized roofing sheets) Landward relocation of buildings
Reduced fish catch[a–g]	*Intensification strategies* Spending longer times at sea Use of small meshed nets (e.g., *médokpokonou and* "wan" practices in Benin) Fish trapping (e.g., the *Acadja* practice in Benin) Change of gear and fish species harvested Increased frequency of fishing in lagoons Extended major fishing season *Extensification strategies* Going further out into the sea Fishing in neighboring coastal communities during taboo days in own community (practiced in some beach-seining communities in Ghana) Temporal migration *Diversification/occupational mobility strategies* Taking wage labor jobs during lean seasons Adding supplementary livelihoods like livestock rearing Sand winning (extraction) Planting, harvesting, and sale of coconut fruits *Storage strategies* Preservation of fish for sale in lean seasons and/or to sell in new and fairer markets Reducing household expenditure in order to save money Borrowing from money lenders Remittances *Experimentation strategy* Varying the hours within the day when fishing expeditions are made
Encroachment of industrial trawlers into artisanal fishing zone	*Experimentation strategy* Studying to precede trawlers or waiting a long while after they have gone

[a]Freduah et al. (2018)
[b]Béné (2006)
[c]Perry and Sumaila (2007)
[d]Dossou and Glehouenou-Dossou (2007)
[e]Niyonkuru and Lalèyè (2010)
[f]Daudu et al. (2020)
[g]Adelekan and Fregene (2015)

malaria. Intensified fishing efforts in response to declining catches are the main reasons behind accusations of SSF as being overexploitative and unsustainable.

Adaption studies from Ghana show that SSF communities also extend efforts by going further into the sea, fishing in areas outside those of their own fishing communities, or moving temporarily to other towns or even countries to fish (Perry and Sumaila 2007; Freduah et al. 2018). Some small-scale fishers maintain contact with fishers from other communities through mobile phone communication to share information about the location of big catches. The fishers respond quickly by moving to that location to fish for a day or two. It is common among West African coastal fishing communities to have indigenously instituted no-fishing (taboo) days (Adjei and Sika-Bright 2019). By way of extensification, some fishing crews move from their communities on taboo days to fish in nearby communities as a strategy to increase catch and thus their weekly earnings. These strategies thrive on good social networks and camaraderie between communities.

Adaptations that are based on diversification of livelihood in SSF communities involve either taking on additional sources of income or changing and acquiring different types of fishing gears or both. Daudu et al. (2020) have shown that gear and occupational diversification are priority strategies for some fishing communities in Nigeria. Some members of a fishing household may take up wage labor jobs as complementary livelihood sources all year round or during some parts of the year. However, for households that have temporary non-fishing livelihoods, fishing is an activity to which they always return. Few artisanal fishers ever leave fishing permanently. Fishing, for them, is not just a livelihood activity; it is their culture, an important part of their identity and something they take pride in. West African coastal fishing communities do not have a strong tradition of farming as an additional source of income. However, recent increase in demand of specific products has provided opportunities for the diversification of livelihoods into harvesting local products. For example, in Ghana, the rise of demand for coconut fruits in large cities has created opportunity for fishers living in coastal areas to harvest and sell coconuts from coconut trees in their vicinity. Others harvest from trees growing on "no man's lands." This has become an important complementary source of income for fishing communities. In some instances, community members might engage in illegal activities to diversify sources of income. Sand winning (extraction of sand from beach areas) is a prohibited activity in most communities and illegal in accordance with national policy in most West African countries. However, some members of SSF communities are involved with sand winning for income.

Evolution of storage strategies is commonly reported as an autonomous adaptation in lagoon communities along coastal West Africa. Storage strategies have been identified in fishing communities in Cote d'Ivoire (Njifonjou et al. 2006), Ghana (Freduah et al. 2018), and Nigeria (Adelekan and Fregene 2015). Storage strategies are usually in the form of securing household sustenance by processing (commonly by smoking or drying) and storing fish to be sold in the lean fishing seasons, borrowing from credit lenders, and/or saving money during the major fishing seasons. This money-saving strategy differs from conventional money-saving practices that aim at saving disposable income. It is an active and often necessary saving against lean seasons, unexpected periods like stormy conditions, or close fishing seasons. It has also been widely observed that some rural households are cushioned by remittances from household members who have migrated either temporarily or permanently from their original community. While this may be true for some communities, it may be a largely missing income stream in other communities. For example, Atti-Mama (1998) reported in a study of nine communities around Lake Nokoue in Benin that more than 95% of households did not receive any remittances.

In response to the encroachment of large trawlers into local fishing waters, coastal fishing communities have no clear strategy. As this is a longstanding but still evolving problem, identified adaptive strategies are, at best, experimental. This may be more conveniently termed a coping mechanism rather than an adaptive strategy. Some authors have particularly highlighted that adaptation strategies are different from coping strategies noting that the former are permanent and long-term changes in practices, while the later are temporary and short-term changes in livelihood practices (Scoones 1998).

The Muni Lagoon and Akosua Lagoon community are used as a case study to illustrate the importance of understanding the vulnerability of an SSF community when planning adaptations.

Understanding Vulnerability for Effective Adaptation Planning: Muni Lagoon Case Study

Internal factors can differentiate the experience of vulnerability in fishing communities. This phenomenon presents both challenges and opportunities for effective adaptation planning in fishing communities. A case study of Akosua village, a lagoon fishing community in Winneba, Ghana, demonstrates this (Koomson et al. 2020).

Akosua village is a small migrant fishing community located on the central coast of Ghana on the outskirts of the town of Winneba (Fig. 1). The community is situated on the sandy barrier that separates the Muni Lagoon from the Atlantic Ocean. The lagoon has been well studied as it was designated as a Ramsar (protected wetland) site in 1992 (GEF 1992). The economy of the community is based predominately on beach seining. This is a fishing method that involves the deployment of seine nets into the sea and hauling them back to shore by hand. Beach seining is done mainly by men. The women in the community are the fishmongers, mostly engaged in the processing and selling of fish. In addition to the predominantly fishing-based livelihood, some women engage in petty trading, and some men engage in daily wage labor jobs within the town of Winneba. Farming and livestock rearing are uncommon in the community due to the sandy and saline nature of the land. The poor quality of the land for farming is common to all lagoon communities and limits opportunities for alternative livelihoods. To understand how vulnerability is manifested within the community, the structure of the fishing activity must be appreciated. Fishing crews are typically made up of the "net owner" (i.e., the owner of the boat, outboard motor, and seine net), boat crew (the net owner's permanent employees), and net draggers (casual workers who haul the net to shore). This crew hierarchy delineates households' economic class as profit is typically shared in the ratio 4:2:1 among the net owner, boat crew, and net draggers, respectively. Crews range from about 12 to 35 men depending on the size of boat and net. The major fishing season starts from the end of the major rainy season (July/

August) to its beginning (March/April) the following year. The period in between (April–July) is the major rainy season characterized by stormy conditions. Thus, no or very limited fishing activities take place during that part of the year. The community major fishing seasons are considered the "good season" when catches are big and incomes are high and the "bad season" when catches are small and incomes are low. This is purely subjective but reflects the community's perception of their well-being during the year. Besides the interruption of fishing activities and income flows by unpredictable stormy conditions, sudden rainfall is a major problem to fish processing. The major fish processing method used in the community is drying in the open fields (Fig. 2). Unexpected rainfall, therefore, results in losses as the half-dried fish in the open fields deteriorate upon contact with water. Besides this, long periods of dryness cause the surface area of the lagoon to decrease in size and reduce the lagoon's water quality. Changes to the lagoon size and water quality have knock-on detrimental effects on fish productivity and species diversity.

The study integrated the concepts of vulnerability and the capital assets component of the sustainable livelihoods approach (SLA) to understand how the lives and livelihood of the community are impacted by climate variability. Drawing upon the IPCC's definition of vulnerability as a function of the magnitude of climate variation to which a system is exposed (E), its sensitivity (S), and its adaptive capacity (AC), a simple equation that multiplies E, S, and AC was used to quantitatively estimate the vulnerability of each household. This is a commonly used methodology according to the Intergovernmental Panel on Climate Change (IPCC 2014). The SLA is a framework designed to identify combinations of livelihood resources, which give people the ability to adopt livelihood strategies that produce their desired outcomes, as well as the institutional processes that mediate their ability to carry out such livelihood strategies within defined policy and ecological and socioeconomic

Fig. 2 Drying fish on a dry lagoon bed in Winneba, Ghana. (Photo D. Koomson)

contexts (Scoones 1998). At the heart of the SLA is the concept that rural communities draw upon and combine five main capital assets (natural, physical, social, human, and financial) to build their livelihoods and respond to shocks. An assessment of the levels of these capital assets and barriers to their access gives insight into what policies are necessary to expand the livelihood platform, thereby improving adaptive capacity and reducing vulnerability (Freduah et al. 2018).

Understanding the experience of vulnerability and autonomous adaptations to climate change in the Akosua community, and the role that internal factors play, has implications for adaptation planning in the community and other communities with similar contexts along the West African coastline. The scale of economic impact that bad seasons have on the household incomes is very important in determining vulnerability. Comparing incomes in good and bad seasons shows that household income drops by about 90% in a bad season (Koomson et al. 2020). With good seasons becoming rare with increasing climate change, food security is threatened. Vulnerability to impacts is not felt in the same way across the community. It is differentiated depending on the gender and wealth status of a household head. For example, in bad seasons, a household in a lower wealth rank and headed by a woman may be comparably less vulnerable than a household of a higher wealth rank and headed by a man. Thus, the most vulnerable are not necessarily female-led and poorer households, as might be expected. Adaptation planning, such as the introduction of alternative/complementary livelihoods, therefore, must be flexible enough to respond to the cyclical and yet unpredictable nature of bad seasons and prioritize the most vulnerable. Such an intersectional understanding of vulnerability is essential for effective adaptation planning because the factors that are responsible for determining who is more vulnerable in the community differ depending on how households are categorized for study (Koomson et al. 2020). An intersectional understanding means that each household's experience of impact should be understood based on the different social identities that characterize that household. For example, a poor household, headed by a woman who is a migrant, may experience impacts differently from a poor household, headed by a woman who is indigenous to that place. Achieving optimum success in adaptation implementation would require that important nuances in differential vulnerability, as identified through an intersectional lens, be leveraged to prioritize critically vulnerable households. This detailed understanding of vulnerability in SSF communities is key to informing interventions that support successful adaptations to climate change. The development, implementation, and inaction of adaptation intervention strategies to address climate change in SSF communities can be realized on the large scale by embedding concepts of adaptation and adaptive capacity in regional and national coastal governance and policy.

Policy Implementation and Participatory Governance

Along with the community-level adaptation, there are policy and governance structures that aim to implement sustainable adaptive capacities for SSF communities. Coastal policies create strategies and establish governance processes and mechanisms to enable coastal fishing communities to implement adaptations to protect and deal with climate change. Well-informed and wide-ranging coastal policies acknowledge that subsistence and resource-dependent SSF communities are faced with climate change impacts and need to undertake sustainable practices for their livelihood, well-being, and development. There are a number of overarching international frameworks that influence and guide regional and national coastal governance and policy in West Africa. These directly or indirectly inform strategies for and governance of adaptation to climate change in lagoon communities.

International Policies and Frameworks Influencing Coastal Governance and Policy

At the global level, the Intergovernmental Panel on Climate Change (IPCC), Sustainable Development Goals (SDGs), and Ramsar Convention provide overarching frameworks for considering coastal governance. In addition, the concept of integrated coastal zone management establishes guidance on managing coastlines holistically. The IPCC highlights the need to explicitly address coastal communities in any climate change policy. The SDGs, specifically 1, 6, 7, 12, 14, and 15, draw attention to the threats posed by climate change in coastal communities in countries of the Global South (UN 2015). Sustainable Development Goal 14, "Conserve and sustainably use the oceans, seas and marine resources for sustainable development," is particularly pertinent for the resources, economy, and livelihood of these coastal communities (UN 2015). This is manifested through the vision of the "Blue Economy": the sustainable use of ocean resources for economic growth, improved livelihoods and jobs, and ocean ecosystem health (World Bank 2015). The concept of integrated coastal zone management (ICZM) arose from the Earth Summit (Rio Convention) in 1992. The aim of the ICZM was to enable governments to create national agendas to understand and find solutions to managing the complex and dynamic nature of the coastal environment. The ICZM concept is embedded in the ecosystem approach, a strategy for the integrated management of land, water, and living resources that promotes conservation and sustainable use in an equitable way. Both the ICZM and the ecosystem approach require that management of the coastline incorporates the impact on and implications for coastal communities of resource usage, pollution, livelihoods, and hazards (including climate change). The aim was to merge physical and environmental aspects of coastal management with human needs, with the ultimate benefit of maintaining coastal ecosystems and environments. However, embedding both the ICZM and the ecosystem approach in decision-making and implementation in the Global South, including West Africa, has not proved to be successful, and considerable challenges remain. The International Convention on Wetlands of 1971, otherwise known as the Ramsar Convention, was established as an international cooperation with national action for the conservation and wise use of resources in wetlands (both coastal and inland) of global significance, including coastal lagoons. Of the countries discussed here, Ghana became a member state of the Ramsar Convention in 1988, closely followed by

Togo (1995) and Cote d'Ivoire (1996). Benin and Nigeria became contracting parties in 2000 and 2001, respectively. The ways in which the Convention mandates are implemented by respective governments differ vastly as the Ramsar Convention gives national governments a broad scope for implementation that relies on effective, dynamic and multilevel processes of implementation (Mauerhofer et al. 2015). The Convention does provide a platform for knowledge transfer as well as management oversight and some level of resources for the protection of specific coastal lagoons.

Along with these international conventions, frameworks, and treaties, there have been regional collaborative initiatives relevant to coastal fishing communities. These regional partnerships include measures to sustainably manage marine and coastal resources, including fish stock and coastal habitats, which help augment the resilience of coastal communities to climate change (FAO 2018).

Regional Frameworks and Governance

At the regional level, West African countries have undertaken different collaborative initiatives and frameworks related to aspects of coastal management and governance. These include the West Africa Coastal Areas Resilience Investment Project (WACA ResIP), the Guinea Current Large Marine Ecosystem (GCLME) program (GCLME 2006), the Fisheries Committee for the West Central Gulf of Guinea (FCWC), and the West Africa Coastal Areas (WACA) Management Program.

The West African countries of Benin, Cote d'Ivoire, Togo, and Senegal are part of a six-country West Africa Coastal Areas Resilience Investment Project (WACA ResIP) aimed at strengthening coastal communities in West Africa. This program assesses coastal damages in terms of economic impacts, environmental implications, and social effects of changing climate. Since 2005, the joint United Nations Development Programme-Global Environment Facility-United Nations Industrial Development Organization (UNDP-GEF-UNIDO) regional collaboration known as the Guinea Current Large Marine Ecosystem (GCLME) program has existed to reduce poverty and enhance economic development of coastal communities through the provision of country-driven mechanisms to support marine and coastal goods and services. It comprises 16 countries of West Africa, including Cote d'Ivoire, Ghana, Benin, Togo, and Nigeria. The GCLME has its roots in the late 1990s when the ecosystem approach was introduced to assess and manage marine resources in West Africa. It provided a framework to promote the ecosystems approach and broaden the scope of coastal management in the region. It prioritized five aspects of coastal communities and environments: (i) productivity, (ii) fish and fisheries, (iii) pollution and ecosystem health, (iv) socioeconomic conditions, and (v) governance. The GCLME enabled large investments to combat coastal change and implement management strategies for coastal issues in the region. An emphasis was placed on monitoring physical processes and hard engineering solutions with little recognition of lived experience of coastal communities. Coupled with weak assessment and governance and an inadequate understanding of the complexities facing coastal communities, the programs put in place through the GCLME have not been successful.

The FCWC and WACA have focused on small-scale fishing. Goals of these programs have been to promote sustainable fisheries management, build strong fisher folk associations, support women in fisheries, ensure efficient fisheries value chains, and identify the challenges of overfishing and other competing uses for coastal areas. The Fisheries Committee for the West Central Gulf of Guinea (FCWC) is a regional integration plan for the Food and Agriculture Organization to ensure the "the blue economy" of the region is based on the participation of small-scale fishers in resource management. It was created by the nations of Benin, Ghana, Togo, Nigeria, Liberia, and Cote d'Ivoire to boost the fisheries sector, by ensuring the sustainable management of their fisheries resources for export and small-scale fishing communities. The West Africa Coastal Areas (WACA) Management Program was launched in 2018 by the World Bank. At its center, the WACA provided access to knowledge and technical expertise, made political and financial commitments, and supported a coordinated, regional approach for coastal resilience in West Africa to enhance adaptive capacity of fishing communities. While these multiple regional and international efforts have been well-intentioned, they suffer from some major drawbacks. Multiple initiatives have led to overlapping governance efforts, with complex organizational structures, funding, and implementation processes that have hampered the success of the initiatives. A top-down approach with limited local input and ownership of projects also meant that the long-term success of initiatives was restricted (Wiafe et al. 2013). All of these initiatives and frameworks emphasized management of the coastline and near-shore environments rather than focusing on climate change. The National Adaptation Programme of Action (NAPA) plans do specifically address adaptations to climate change in countries of the Global South and inform climate change policy relevant to coastal environments and communities.

National-Level Coastal and Climate Change Policy Mechanisms

The National Adaptation Programme of Action (NAPA) is national-level adaptation plan submitted to the United Nations Framework Convention on Climate Change (UNFCCC) secretariat by the least developed member states in 2007 and 2008. These documents outline the threat to individual nations and their planned priority adaptation actions. Development of the Intended Nationally Determined Contributions (INDC) followed, submitted in 2015. The INDC outline the priority adaptation sectors and interventions for a country and run from 2016 to 2030. The National Adaptation Programme of Action (NAPA) for coastal West African countries acknowledged the adverse impacts of climate change on SSF communities, signposting threats to these communities and their livelihoods and creating adaptation plans (Table 2).

The selection of priority adaptation interventions for NAPAs (and INDCs) was based on a multi-criteria analysis that included monetary cost and benefit, potential economic benefits and synergy with other environmental policies, as well as potential contribution to poverty alleviation. While it is noteworthy that SSF communities and their livelihood are quite significantly acknowledged in national policies, none

Table 2 Climate change adaptation plans for coastal SSF communities across West Africa

Country	Stated adaptation plans for SSF communities
Ghana[a,b]	Enhance fisheries resource management Develop alternative sources of livelihoods for fisher folk Promote fish farming
Nigeria[c,d]	Enhance artisanal fisheries and encourage sustainable aquaculture Encourage the provision of social welfare programs by faith-based and civil society organizations to address the climate change-induced needs of vulnerable groups
Togo[e]	Protect the coastal zone Strengthen infrastructure resistance to climate change in coastal zones Strengthen community resilience to climate change in the coastal zone
Benin[f]	Ensure the protection of the shoreline against the risk of sea level rise Reduce the vulnerability of human settlements and resources located in the coastal area to sea level rise Ensure continuously the protection of marine and lagoon ecosystems
Cote d'Ivoire	All policy documents for this country are in French (official language of the country) thus not accessible for review purposes

[a]Ghana's INDC (2015)
[b]National Climate Change Adaptation Strategy (2010–2020)
[c]Nigeria's INDC (2015)
[d]National Adaptation Strategy and Plan of Action on Climate Change for Nigeria (NASPA-CCN, 2011)
[e]Togo's INDC (2015)
[f]Readiness Project for Benin's Intended Nationally Determined Contributions (2018)

of these actions made it into the priority actions for any of the countries, which were mostly focused on agriculture, transportation, energy, and health sectors. The MCA approach is likely responsible for the prioritization of other livelihoods like rural agriculture over SSF fishing. As is widely noted in the SSF literature, the contribution of SSF to national economies, in terms of GDP, is relatively small. Poverty alleviation, food security, and rural economic development are the most significant contributions of SSF. However, it is also acknowledged that in comparison to other rural livelihood activities like farming, SSF has higher exposure and sensitivity to climate change impacts. While the MCA process is not completely described in the reviewed documents, it could be conjectured that if the disproportionately higher vulnerability of SSF communities had been considered and higher weights assigned to poverty alleviation and food security, adaptation interventions for SSF communities would have risen in priority. Countries differed in which aspects of coastal adaptations took precedent (Table 2). While some countries (Ghana and Nigeria) focused on livelihoods of coastal communities by establishing means for improving and enhancing income generations, others such as Benin, Togo, and Cote d'Ivoire concentrated more on hard engineering solutions to protect coastal infrastructure (Table 2). The NAPA were not as wide-ranging or successful as they might have been at addressing coastal adaptations to climate change.

All countries adopted a multi-sectoral approach to adaptation planning and demonstrate an appreciation of the need to connect adaptation planning with the broader national development agenda. However, the multifaceted and complex interlinkages between climate change, natural resources, and local development issues that underpin vulnerability of a community, and which are critical to successful adaptation interventions, were not fully appraised as part of the plans. In addition, NAPA implementation required strong cross-sectoral collaboration to realize the planned adaptation interventions, and this was not necessarily in place due to gaps between national policy and local implementation.

Adaptation to Climate Change in Coastal Communities: The Policy Gap

Institutional arrangements in Africa have conventionally followed a formal, top-down framework. The national and regional policy agreements in Benin, Nigeria, Ghana, Togo, and Cote d'Ivoire aimed at coastal management and enhancing adaptive capacity of coastal communities are managed and implemented by government organizations, often in partnership with nongovernmental organizations. This multi-organizational, top-down approach has resulted in several barriers to the successful implementation of policy-driven climate change adaptation strategies at the coast. Policy deployment through multiple national and nongovernmental agencies can often result in uncoordinated and overlapping strategies resulting in confused implementation at the local level and limited success of adaptation strategies. The current national governance top-down approach allows limited opportunities for local governance. This creates a gap, "the policy gap," between the well-intentioned policy discourse at the national government level and the actual implementation of policy at the village level. The top-down approach also results in failure to understand and address the highly localized and often community-specific conditions required to enact successful climate change adaptations. When multi-institutional governance structures include engagement opportunities for local communities, the success of coastal management plans and adaptation strategies is greatly enhanced. For example, an understanding of specific local environmental conditions and socioeconomic responses is needed when developing adaptations to sea level rise, but national policies currently do not support that local oversight (Hauer et al. 2020). In addition to the difficulties created by pursuing a top-down approach, the policies themselves have tended to have a disproportionate focus on physical and resource management of the coastline rather than adaptations that address the social and human impacts of climate change (Clarke et al. 2013; Butler et al. 2015; Cinner and Barnes 2019). The physical effects of climate change have been prioritized over livelihood, well-being, and cultural needs of communities. Such an approach requires reconciling lagoon conservation issues with the livelihood interests and local economy of lagoon communities. In Ghana, an example of the disconnect between national coastal resource policy and local community needs is seen in the mismanagement of mangroves. Mangroves are overexploited by lagoon communities, despite playing a vital role in the physical protection of coastal areas, and mangrove conservation plans at lagoons have suffered from a lack of administrative coordination across ministries and agencies (Boateng 2018). Conventional multi-institutional, top-down systems of government have not been very successful in resolving or managing coastal problems (Clarke et al. 2013). The need to shift to a human-centered, bottom-up approach to managing lagoon environments and their communities, where communities are participants to the planning, co-production, and implementation of coastal adaptations to climate change, is key to their long-term resilience and sustainability (Butler et al. 2015).

Participatory Governance

The recognition of the gap between national policy discourse and local implementation has prompted a shift to the development of a more participatory approach that ensures coastal communities are central to the joint management of their resources and local environment. In this participatory approach, community knowledge is used to bridge the gap between national policy, climate research, and lived experience of climate change (Clarke et al. 2013). The resultant integrated governance systems to support climate change planning and adaptations combine local participatory governance with multi-institutional government and nongovernmental organizations. This bottom-up approach ensures that community needs are addressed and local conditions are taken into account in adaptations to climate change and the management strategies that support them. In effect, a balance between "top-down" policy implementation and "bottom-up" community-driven needs that maps the top-down agenda to a community-based approach is recommended for successful adaptive strategies (Butler et al. 2015). In this approach, there is a recognition of the contextual factors through local stakeholder engagement as well as the integration of indigenous knowledge and local approaches to understand, manage, and create solutions for climate change adaptations (Cinner and Barnes 2019). Indigenous knowledge and traditional practices have developed over generations by communities coping with climatic variability and extreme weather (FAO 2018), such as in Nigeria, where SSF communities use their traditional knowledge of local meteorological variation to inform adaptations to the management of coastal flooding.

Most of the research about participatory governance of coastlines in West Africa focuses on Ghana and Nigeria with fewer studies from Benin, Togo, and Cote d'Ivoire. Ghana has many agencies with a top-down institutional framework, including those that govern the coastal and marine environment. However, progress is being made toward creating participatory structures for coastal management and climate change adaptation, specifically around the areas of resource management, environmental monitoring, and community education (Wiafe et al. 2013). The participatory approach is also being used in fishing communities along the Badagry coast of Nigeria to create and implement local adaptations to climate and environmental change (Olawepo 2008). Another example is from the Aby Lagoon in Côte d'Ivoire, where co-management of fishing resources has given fishing communities a sense of engagement, ownership, and empowerment that has supported sustainable livelihoods, enhancing resilience to climate change (Njifonjou et al. 2006). The success of these participatory approaches is encouraging for future implementation of climate change adaptations in coastal lagoon communities across West Africa.

Summary

The coastlines of Cote d'Ivoire, Ghana, Benin, Togo, and Nigeria are dominated by dynamic lagoon environments. Sea level rise and climatic changes are affecting those lagoons and their associated nearshore environments. The fishing communities that live next to lagoons depend on them for resources and ecosystem services that support economic livelihood and well-being. The vulnerability of lagoon communities to climate change is related not only to the physical and environmental effects of climate change but also to local resource use, sociocultural community structures, susceptibility to external factors, population growth, and governance infrastructures. The interconnected nature of climate-related challenges with development issues complicates the success of climate change adaptations and interventions. Adaptations enacted as part of a managed intervention plan require an understanding that vulnerability to climate change is not felt in the same way in different communities nor even across a single community and an acknowledgment of intersectionality is required when deciding how and where to place adaptation interventions.

Multiple frameworks exist in West Africa that are related to the management of the coastal zone, and National Adaptation Programme of Action (NAPA) plans in the region have included coastal communities. National coastal policy arising from such frameworks should be the vehicle for managing climate change adaptations in lagoon communities. However, where such policy exists, it has suffered from formal top-down approaches with gaps between policy and its implementation as well as uncoordinated approaches. Importantly, the policies themselves have failed to acknowledge the human experience of climate change, tending to focus on physical change at the coast. To rectify this, climate and coastal change policy must evolve to embed concepts of adaptation and adaptive capacity, giving the same consideration to community livelihood and well-being as to the physical environment. An integrated participatory approach, which allows lagoon communities to play a key role in the governance of the coastal zone, will enable such a human-centric approach and facilitate the integration of climate change adaptation into coastal planning and management.

References

Adelekan IO (2010) Vulnerability of poor urban coastal communities to flooding in Lagos, Nigeria. Environ Urban 22(2):433–450. https://doi.org/10.1177/0956247810380141

Adeniyi MO (2016) The consequences of the IPCC AR5 RCPs 4.5 and 8.5 climate change scenarios on precipitation in West Africa. Clim Chang 139:245–263. https://doi.org/10.1007/s10584-016-17774-2

Adelekan I, Fregene T (2015) Vulnerability of artisanal fishing communities to flood risks in coastal southwest Nigeria. Clim Dev 7(4):322–338. https://doi.org/10.1080/17565529.2014.951011

Adjei JK, Sika-Bright S (2019) Traditional beliefs and sea fishing in selected coastal communities in the Western Region of Ghana. Ghana J Geogr 11(1):1–19. https://doi.org/10.4314/gjg.v11i1.1

Agrawal A (2008) The role of local institutions in adaptation to climate change. World Bank. https://doi.org/10.1596/28274

Akanni KA (2010) Fishing technologies and catch levels among small-scale fishers in Lagos state, Nigeria. N Am J Fish Manag 30:309–315. https://doi.org/10.1577/M08-061.1

Anthony EJ, Oyede LM, Lang J (2002) Sedimentation in a fluvially infilling, barrier-bound estuary on a wave-dominated, microtidal coast: the Ouémé River estuary, Benin, west Africa. Sedimentology 2002 49(5):1095–1112. https://doi.org/10.1046/j.1365-3091.2002.00491.x

Appeaning-Addo K (2014) Managing shoreline change under increasing sea-level rise in Ghana. Coast Manag 42:555–567. https://doi.org/10.1080/08920753.964820

Atti-Mama C (1998) Co-management in continental fishing in Benin: the case of lake Nokoué. In The International Workshop on Fisheries Co-Management. (NO DOI)

Awoskia LF, Ibe AC (1993) Geomorphology and tourism related aspects of the Lekki barrier-lagoon coastline in Nigeria. In: Wong PP (ed) Tourism vs environment: the case for coastal areas. The geo-journal library v. 26. Springer, Dordrecht. https://doi.org/10.1007/978-94-011-2068-5_9

Béné C (2006) Small-scale fisheries: assessing their contribution to rural livelihoods in developing countries. FAO Fish Circ 1008:46. (NO DOI)

Bernacsek GM, Hughes JS, Hughes RH (1990) A directory of African wetlands. (NO DOI)

Blankespoor B, Dasgupta S, Laplant B (2014) Sea level rise and coastal wetlands, Ambio. The Royal Swedish Academy of Sciences. https://doi.org/10.1007/s13280-014-0500-4

Boateng I (2012) An application of GIS and coastal geomorphology for large scale assessment of coastal erosion and management: a case study of Ghana. Journal of coastal conservation 16 (3):383–397. https://doi.org/10.1007/s11852-012-0209-0

Boateng I (2018) An assessment of vulnerability and adaptation of coastal mangroves of West Africa in the face of climate change. In: Makowski C, Finki CW (eds) Threats to mangrove forests, coastal research library 25. https://doi.org/10.1007/978-3-319-73016.5_7

Brown S, Kebede AS, Nicholls RJ (2011) Sea level rise and impacts in Africa 2000–2010. http://www.unep.org/climatechange/adaptation/Portals/133/documents. (NO DOI)

Butler JRA, Wise RM, Skewes TD, Bohensky EL, Peterson N, Suadnya W, Yanuartati Y, Handayani T, Habibi P, Puspadi K, Bou N, Vaghelo D, Rochester W (2015) Integrating top-down and bottom-up adaptation planning to build adaptive capacity: a structured learning approach. Coast Manag 43:346–364. https://doi.org/10.1080/08920753.2015.1046802

Cinner JE, Barnes ML (2019) Social dimensions of resilience in social-ecological systems. One Earth 1:51–56. https://doi.org/10.1016/j.oneear.2019.08.003

Chambers R, Conway G (1992) Sustainable rural livelihoods: practical concepts for the 21st century. Institute of Development Studies (UK). https://opendocs.ids.ac.uk/opendocs/handle/20.500.12413/775

Clarke B, Stocker L, Coffey B, Leith P, Harvey N, Baldwin C, Baxter T, Bruekers G, Galano CD, Good M, Haward M, Hofmeester C, De Freitas DM, Mumford T, Nursey-Bray M, Kriwoken L, Shaw J, Shaw J, Smith T, Thomsen D, Wood D, Cannard T (2013) Enhancing the knowledge-governance interface: coasts, climate and collaboration. Ocean Coast Manag 86:88–99. https://doi.org/10.1016/j.ocecoaman.2013.02.009

Danladi IB, Kore BM, Gul M (2017) Vulnerability of the Nigerian coast: an insight into sea level rise owing to climate change and anthropogenic activities. J Afr Earth Sci. https://doi.org/10.1016/j.jafrearsci.2017.07.019

Davies-Vollum KS, West M (2015) Shoreline change and sea level rise at the Muni-Pomadze coastal wetlands (Ramsar site), Ghana. J Coast Conser Manage 19(4):515–525. https://doi.org/10.1007/s11852-015-0403-y

Daudu AK, Oladipo FO, Olatinwo LK, Kareem WO (2020) Analysis of factors influencing vulnerability of fishing systems to climate change among Artisanal Fisher-folks in Coastal area of Lagos, Nigeria. Afr J Land Policy Geospat Sci 3(1):80–98. Retrieved from https://revues.imist.ma/index.php/AJLP-GS/article/view/17924 (No DOI)

Dey MM, Gosh K, Valmonte-Santos R, Rosegrant MW, Chen OL (2016) Economic impact of climate change and climate change adaptation strategies for fisheries sector in Fiji. Mar Policy 67:164–170. https://doi.org/10.1016/j.marpol.2015.12.023

Djihouessi BD, Djihouessi MB, Aina MP (2016) A review of habitat and diversity research in Lake Nokue, Benin Republic: current state of knowledge and prospects for future research. Ecohydrol Hydrobiol 19:131–145. https://doi.org/10.1016/j.ecohyd.2018.04.003

Diop S, Sherin PA (2016) Sustainable oceans and coasts: lessons learnt from eastern and Western Africa. Estuar Coast Shelf Sci 18:327–339. https://doi.org/10.1016/j.ecss.2016.03.032

Dossou KM, Glehouenou-Dossou B (2007) The vulnerability to climate change of Cotonou (Benin) the rise in sea level. Environ Urban 19(1):65–79. https://doi.org/10.1007/978-94-007-7088-1_25

Druyan L (2011) Review: studies of 21-century precipitation trends over West Africa. Int J Climatol 31:1415–1424. https://doi.org/10.1002/joc.2180

Duck RW, de Silva JF (2012) Coastal lagoons and their evolution. Estuar Coastal Shelf Sci 110:2–14. https://doi.org/10.1016/j.ecss.2012.03.00

Evadzi PIK, Scheffran J, Zurita E, Hunicke B (2018) Awareness of sea-level response under climate change on the coast of Ghana. J Coast Conserv 22:183–197. https://doi.org/10.1007/s11852-017-0569-6

FAO (2018) The state of world fisheries and aquaculture 2018 – meeting the sustainable development goals. FAO. (No DOI)

Freduah G, Fidelman P, Smith TF (2018) Mobilising adaptive capacity to multiple stressors: insights from small-scale coastal fisheries in the Western Region of Ghana. Geoforum 91:61–72. https://doi.org/10.1016/j.geoforum.2018.02.026

GCLME (2006) Guinea current large marine ecosystem (GCLME)- transboundary diagnostic analysis: a programme of the governments of the GCLME countries under GEF/UNIDO/UNDP/UNEP/US-NOAA/NEPAD/FAO and IMO. (No DOI)

GEF (1992) Coastal Wetlands Management Project, Republic of Ghana. Global Environment Facility. Project document, August 1992. World Bank. (No DOI)

Ghana Embassy (2014) Population. http://www.ghanaembassy.org/index.php?page=population. (No DOI)

Hastings RA, Rutterford LA, Freer JJ, Collins RA, Simpson SD, Genner MJ (2020) Climate change drives poleward increases and equatorward declines in marine species. Current Biology 30(8): 1572–1577.e2

Hauer ME, Fussell E, Mueller V et al (2019) Sea level rise and human migration. Nat Rev Earth Environ 1:28–39. https://doi.org/10.1038/s43017-019-0002-9

IPCC (2014) In: Barros VR, Field CB, Dokken DJ, Mastrandrea MD, Mach KJ, Bilir TE, Chatterjee M, Ebi KL, Estrada YO, Genova RC, Girma B, Kissel ES, Levy AN, MacCracken S, Mastrandrea PR, White LL (eds) Climate change 2014: impacts, adaptation, and vulnerability. Part B: regional aspects. Contribution of working group II to the fifth assessment report of the intergovernmental panel on climate change. Cambridge University Press, Cambridge, UK/New York, p 688. (No DOI)

Koomson D, Davies-Vollum KS, Raha D (2020) Characterising the vulnerability of fishing households to climate and environmental change: insights from Ghana. Mar Policy 120 (2020):104142. https://doi.org/10.1016/j.marpol.2020.104142

Lam VW, Cheung WW, Reygondeau G, Sumaila UR (2016) Projected change in global fisheries revenues under climate change. Sci Rep 6:32607. https://doi.org/10.1038/srep32607

Mauerhofer V, Kim RE, Stevens C (2015) When implementation works: a comparison of Ramsar convention implementation in different continents. Environ Sci Pol 51:95–105. https://doi.org/10.1016/j.envsci.2015.03.016

Njifonjou O, Satia B, Angaman K (2006) Fisheries co-management and poverty alleviation in the context of the sustainable livelihoods approach: a case study in the fishing communities of aby lagoon in cote d'Ivoire. Int J Sustain Dev World Ecol 13(6):448–458. (No DOI)

Niyonkuru C, Lalèyè PA (2010) Impact of acadja fisheries on fish assemblages in Lake Nokoué, Benin, West Africa. Knowl Manag Aquat Ecosyst 399:05. https://doi.org/10.1051/kmae/2010033

Ochiewo J (2004) Changing fisheries practices and their socioeconomic implications in south coast Kenya. Ocean Coast Manag 47(7):389–408. https://doi.org/10.1016/j.ocecoaman.2004.07.006

Odunga S, Badru G, Olalekan MB (2014) Climate change, sea level rise and coastal inundation along part of Nigeria barrier lagoon coast. J Appl Sci Environ Manage 18(1):41–47. https://doi.org/10.4314/jasem.v18i1.6

Olawepo RA (2008) Using participatory rural appraisal to explore coastal fishing in Badagry villages, Nigeria. Environmentalist 28:108–122. https://doi.org/10.1007/s10669-007-9052-5

Perry RI, Sumaila UR (2007) Marine ecosystem variability and human community responses: the example of Ghana, West Africa. Mar Policy 31(2):125–134. https://doi.org/10.1016/j.marpol.2006.05.011

Rahmstorf S (2007) A semi-empirical approach to projecting Future Sea-level rise. Science 315:368–370. https://doi.org/10.1126/science.1135456

Sagoe-Addy K, Appeaning Addo K (2013) Effect of predicted seal level rise on tourism facilities along Ghana's coast. J Coast Conserv 17:155–166. https://doi.org/10.1007/s11852-012-0227-y

Scoones I (1998) Sustainable rural livelihoods: a framework for analysis. IDS working paper 72. https://opendocs.ids.ac.uk/opendocs/handle/20.500.12413/3390

Short R, Gurung R, Rowcliffe M, Hill N, Milner-Gulland EJ (2018) The use of mosquito nets in fisheries: a global perspective. PLoS One 13(1):e0191519. https://doi.org/10.1371/journal.pone.0191519

USAID (2014) Sustainable wetlands adaptation and mitigation programme: final report. https://www.cifor.org/fileadmin/subsites/twincam/reports/SWAMP_Final_Report_2014.pdf. (No DOI)

UN (2015) Transforming our world, The 2030 Agenda for Sustainable Development. https://sustainabledevelopment.un.org/post2015/transformingourworld/publication. (No DOI)

Wiafe G (2010) Coastal and continental shelf processes in Ghana. NTIS technical report. University of Ghana, Accra. http://www.onr.navy.mil/reports/FY11/lowaife.pdf. (No DOI)

Wiafe G, Boateng I, Appeaning-Addo K (2013) Handbook for coastal processes and management in Ghana. Gloucester: Choir Press, 274 p.

World Bank (2019) West Africa's coast: losing over £3.8 Billion a year to erosion, flooding and pollution. https://www.worldbank.org/en/region/afr/publication/west-africas-coast-losing-over-38-billion-a-year-to-erosion-flooding-and-pollution. (No DOI)

World Bank Group (2015) Economic, environmental, and social evaluation of Africa's small-scale fisheries environment and natural resources global practice policy note

Zacarias DA (2019) Understanding community vulnerability to climate change and variability at a coastal municipality in southern Mozambique. Int J Climate Change Strateg Manage 11(1):154–176. https://doi.org/10.1108/IJCCSM-07-2017-0145

10

Climate Variability and Rural Livelihood Security: Impacts and Implications

Kehinde Olayinka Popoola, Anne Jerneck and Sunday Adesola Ajayi

Contents

Abstract

In a typical Nigerian village, the majority of the population comprises old people who are mostly economically unproductive due to reduced or loss of physical strength brought about by ageing and ill health. Many of these rural old people still work, and do so outside the formal sector, and are particularly susceptible to the effects of climate variability and change. Few studies have reported on climate change and the rural aged and there is a research gap as regards rural elderly peoples' perception of climate variability impact on them. Since little is known

K. O. Popoola (✉)
Department of Urban and Regional Planning, Obafemi Awolowo University, Ile-Ife, Nigeria

A. Jerneck
Lund University Centre for Sustainability Studies, Lund, Sweden
e-mail: anne.jerneck@LUCSUS.lu.se

S. A. Ajayi
Department of Crop Production and Protection, Obafemi Awolowo University, Ile-Ife, Nigeria

Institute for Sustainable Development, First Technical University, Ibadan, Nigeria
e-mail: sajayi@oauife.edu.ng

about their perception of climate variability impacts and implications on the rural aged especially in relation to their livelihood activities in Sub-Saharan Africa, this chapter therefore examined the impact of climate variability on the livelihood security of the rural aged in different ecological zones of Nigeria.

Both qualitative and quantitative methods were used for data collection. Qualitative data were obtained through interviews with four aged and four aged women selected purposively in each rural community and analyzed using Content Analysis Method. Quantitative data were obtained through structured questionnaire administered to an aged male and an aged female population available in selected houses (the aged are people 60 years and over in age) in selected rural communities in selected ecological zones of Nigeria. Where there was no combination of the two (aged men and aged women), either of the two was also sufficient.

It was discovered that the ageds' experiences of climate variability impact relate to the prevailing climate variability characteristic of each ecological zones. The impact on their livelihood in these zones is seen in terms of livestock death, lack of pastures for herds, scarcity of water, pest invasion, delayed planting crop failure, need for irrigation, water logging, drowning of small animals, human and animal illness. This means that planning decisions related to climate change issues should take cognizance of the views of the aged populations especially of those residing in rural areas as they are the most affected by the impact.

Keywords

Climate variability · Rural aged · Livelihood security · Ecological zones

Introduction

Climate change implies a shift in the main state of a climate or in its variability, persisting for an extended period that can be decades or longer (Intergovernmental Panel on Climate Change 2007). It has been known for a while to have major global environmental, social, and economic impact (Scholze et al. 2006; Douglas et al. 2008) and poses a major serious threat to sustainable development with adverse effects on the environment, human health, food security, economic activities, natural resource management, and physical infrastructure (Government of Kenya (GOK) 2010). Climate change poses serious risks to society because of the physical characteristics of the planet, the biological resources on which human lives depend, and current social systems that are adapted to existing rather than changing climate conditions (Higgins 2014).

Climate change is rapidly emerging as one of the most serious global problems affecting many sectors in the world especially livelihood. Negative consequences in Africa are already prevalent in terms of frequent floods, droughts, and shift in marginal agricultural systems (Collier et al. 2008). Deschenes and Greenstone

(2012) explained that climate change affects the two most important direct agricultural production inputs, precipitation, and temperature, which in turn affects livelihood. Studies have shown that Africa is getting more vulnerable to climate change/variability impact on livelihood compared to other parts of the World. This is because the livelihoods for 80% of the African population are dependent on rain-fed agriculture which are vulnerable to multiple stresses such as climate change (World Bank 2008). Countries in sub-Saharan Africa may likely suffer most from these consequences because of their greater reliance on climate-sensitive resources like agriculture for sustenance of livelihoods (Eboh 2009).

In Nigeria, about 60% of the populations are rural residents depending on agriculture for their livelihood (Central Bank of Nigeria 2006) and studies have confirmed serious effects of climate change on this population. Their vulnerability to climate change impacts is largely due to the socioeconomic and political context (Thornton et al. 2014). However, all households within a community are not equally vulnerable to climate change impact (Adger 2003). This is because vulnerability is highly differentiated (O'Brien et al. 2007) and varies across: age, gender, income, and type of livelihood. In Nigeria, women, children, and the elderly will be most vulnerable to climate change impact, especially the rural elderly who are economically less productive due to physical weakness brought about by ageing and ill health (Anele 2012).

Mcleman and Hunter (2010) explained that climate change events, impacts on livelihood and vulnerability differ considerably across geographic and ecological regions. This is because livelihoods are ecologically dependent and some regions and activities are more sensitive to climate change than others (Saarinen et al. 2012). It is therefore relevant to consider ecological characteristics in relation to climate change impacts on livelihood. Also in Sub-Saharan Africa, extreme droughts is already constraining production of food and rearing of livestock (Kebede et al. 2011). For 2050, it is projected that, especially in West Africa, yields could drop by 20–50% due to variability of rainfall reducing length of the growing season (Sarr 2012). There are many studies on climate change/variability and livelihood (Ofoegbu et al. 2017; Ubisi et al. 2017; Aniah et al. 2016; Egbe et al. 2014; Okonkwo et al. 2015; Zeleke and Aberra 2014). For instance, Aniah (2016) examined the effects of climate change on livelihoods of smallholder farmers in the Upper East Region of Ghana. The results revealed that droughts, floods, pest and diseases, postharvest losses, declining crop yields, and animal production have threatened smallholder farmers' livelihood activities. Also, Ubisi et al. (2017) investigated the perceived effects of climate change on crop production and household livelihoods of smallholder farmers in Mopani and Vhembe district, South Africa. Their study showed that subsistence farmers perceived prolonged droughts as an impact of climate change leading to low crop yield and high crop failure. These studies focused on the impact of climate change/variability on livelihood activities but the impacts on the livelihood activities of the aged were not considered. Against this backdrop, we have reason to believe that climate change impact on livelihood among the rural aged is likely to vary with geographic location and livelihood activities. Based on Nigeria as a case, this chapter provides answers to this question: How does the perceived climate variability affect rural livelihood activities of the aged population in selected rural communities of the different ecological zones of Nigeria?

Multistage Selection of Ecological Zones and the Aged

Nigeria, situated in West Africa on Latitudes 4°0 to 14°0′ North and Longitudes 2°2′ and 14°30′ East, is bounded in the west by the Republic of Benin, in the East by Chad and Cameroon, in the North by Niger Republic, in the South by Gulf of Guinea, and in the North-East by Lake Chad (Map 1). Nigeria is a multiethnic and culturally diverse society divided into 36 autonomous states (World Bank Group 2017). Nigeria has two seasons, the dry and the rainy season, and six ecological zones: Mangrove zone, Forest zone, Montane zone, Guinea savannah zone, Sudan savannah zone, and Sahel savannah zone. This chapter is set in four ecological zones with one state representing each zone–Guinea savannah zone of Oyo State (Oyo State), Mangrove zone of Ondo State (Ondo State), Sudan savannah zone (Kebbi State), and Montane zone (Plateau State) (see Map 1).

In the initial stage, the ecological zones (They are: Mangrove zone, Rain forest, Montane region, Guinea savannah, Sudan savannah and Sahel Savannah) were identified based on Keay (1949), Kueppers (1998) classification. Four Ecological Zones were randomly selected out of the six zones. They are: Guinea savannah zone,

Map 1 Map of Nigeria showing the selected states. (Source: Adopted from Popoola and Ajayi 2019)

Mangrove zone, Montane zone, and the Sudan savannah zone. One State was then selected from each of the selected ecological zones. The states are Oyo State in the Guinea savannah zone (Guinea savannah zone of Oyo State), Ondo State in Mangrove zone (Mangrove zone of Ondo State), Plateau State in Montane zone, and Kebbi State in the Sudan savannah zone. In the next stage, two local government areas were selected from each of the States based on their rurality and climate change impact. The local government areas are: Ilaje and Eseodo in Ondo State; Oorelope and Atisbo in Oyo State; Riyom and Kanke in Plateau State; Jega and Yauri in Kebbi State. In the third stage, three rural settlements were selected from each of the local government areas through a simple random selection process. They are: (Ilepete, Obenla, Ebijimi) in Ilaje LGA; (Igbalekagbo, Uropati, Olorunpupa) in Ese-odo LGA; (Alaguntan, ologundudu, Sooro)in Oorelope LGA; (Alakuko, Sabe, Yowere) in AtisboLGA; (Ambob, Tahoss, Kwakwi) in Riyom LGA; (Tse, Jinglay, Lebwit) in Kanke LGA; (Akalawa, Duduke, Gindi) in Jega LGA; (Unguwar Kurya, Zara Bimi, Giwa Tazo) in Yauri LGA. To ensure a fair draw for all settlements each settlement in the respective local government area was assigned a number written on a piece of paper, placed in a container, and then drawn from the container without replacement. In the fourth stage, snowball approach was used to identify the houses where the rural aged resides. The aged were defined as people who are at least 60 years of Age.

Both qualitative and quantitative data were used (Primary Data Source). The quantitative data were obtained through structured questionnaire administered to an aged male and an aged female available in the selected houses in the selected rural communities of the selected ecological zones. Where there were no combinations of the two (aged men and aged women), either of the two was also sufficient. The qualitative data includes observations, key informant interviews, and open as well as semi-structured interviews with four aged men and four aged women selected purposively in each of the selected villages. The interviewer used progressive focusing to ask the participants both new and follow up questions. Detailed field notes and audio recording were made throughout the field observations. Data collected was disaggregated by gender to recognize the different experiences of male and female and analyzed using various methods such as descriptive statistics (frequencies and percentages) and SPSS. In the qualitative analysis, Content Analysis Method was used to identify commonalities in terms of recurring aspects and themes as well as differences and contradictions in terms of deviating views and outliers.

Ageds' Experience of Climate Variability Impact on Their Livelihood Security

In this section findings of climate variability impact on livelihood security of the rural aged in selected ecological zones of Nigeria was presented. In focus, the following domains: agriculture and food security; environment; energy; and water resources; as well as human and animal health; dwellings, transport, and migration were placed. According to Ashwill et al. (2011), climate change impact refers to the social and environmental consequences in terms of water scarcity, loss in agricultural yields, and drops in livestock production, to mention a few. It is important to examine the climate variability impact on livelihood security of the people because it helps informing and tailoring directions for adaptation (Mendelsohn 2008). It is also important to examine the impact on livelihood by gender and geographical locations. Examining the impact by gender helps to understand that the different roles, resources, opportunities, and agency of men and women influence how they experience, perceive, and respond locally to climate change (Ashwill et al. 2011). Vulnerability to climate change impact is highly correlated with its geographical peculiarities (Callo-Concha et al. 2013). Therefore, examining the impact by geographical location helps to determine both the perceived need to adapt and the acceptability of particular adaptive measures because of factors such as: attachment to place, cultural values, social identity, and regional risk attitudes (Adger et al. 2009).

Figure 1 shows the climate variability impact on livelihood security of the rural aged in selected ecological zones of Nigeria. Findings revealed that climate variability has affected the livelihood of the aged to a great extent. For instance, from the table, in Guinea Savannah Zone, 68.3% of the men and 69% of the women complained of water shortage and 60% men and 49.2% women felt the need for irrigation. Also, 42.4% men and 26.2% had to delay their planting; 94.6% of the men and 91.8% of the women lamented about crop failure; 37% of the men and 35.9% of the women worried about their animals getting weak and sick while 32.6% men and 27.9% women had their livestock dead. All these can be attributed to climate

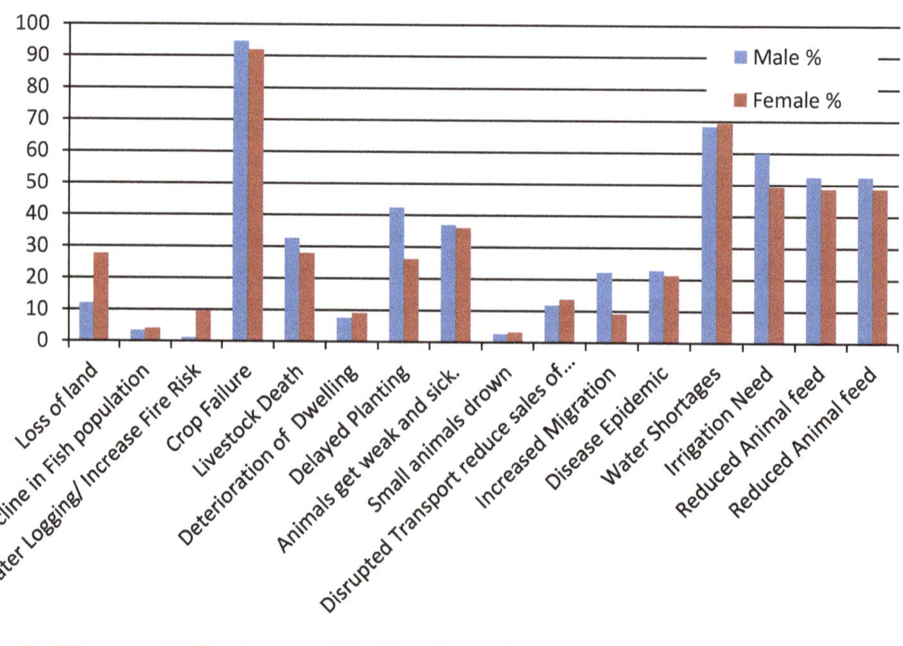

Fig. 1 Climate variability impact of livelihood security of aged men and women in Guinea Savannah Region of Nigeria. (Source: Author's Fieldwork)

variability impact. Below, selected comments from aged men and women sharing their experience of climate variability impact on their livelihood security in the different selected villages of Oorelope and Atisbo Local Government Areas of Oyo State, Guninea Savannah Region of Nigeria were included.

> Before the change in weather, we planted maize twice a year. But now, we plant it only once. If we try to plant the second time, which is usually done in the eight month, the produce will be very small and not profitable. Just like the one I am holding now (respondent holding a sample maize), who will buy this from us? (**Waliu Hamuzat, Male, 64 years**)
>
> Our yam produce is not as big as before again. The presence of yam leaf is a positive sign of a possible bountiful harvest. During the 10th and 11th month of plantation, some cultivated yams will still have their leaves intact, like before, but now, in the 9th month, some yams would have lost their leaves. Infact, the period when tubers had begun to grow proper, that is when we usually experience the lack of rain (drought). And it is usually when the yam needs the rain the most. Most importantly, these are the periods we usually have constant rainfall before. Now the story is different.
>
> Take for example, if I planted about 3000 yam and I was only able to harvest 100. This can only be related to the heat generated from the dry land and I cannot be at peace with such result (**Kazeem Lawal, Male, 65 years**)
>
> There is shortage of food and water, and our cows are going seriously hungry. This has led to a clash between farmers and the Fulani herdsmen because the cows now go to feed on people's farmland. Any cow caught on a farmland, automatically belongs to the farmland owner. In addition, a lot of cows cannot even stand up on their own again due to hunger. The herdsmen at times usually help them up (**Yussuf Dammani, Male, 62 years**)
>
> Fever most especially, and several other ailment usually surface at this period. Since food is what our body needs most, a shortage of it will lead to sickness in the body (**Durodola Lamina, Female, 73 years**)
>
> The little rain we had was more of heavy storm than water. The storm was too much; in fact the structure we erected as mosque was totally destroyed by the storm. It also destroyed a lot of buildings. The storm is a new experience in this area (**Ajao Abebi, Female, 68 years**)

Figure 2 also shows the climate variability impact on livelihood security of the aged men and women in the Coastal Zone of Nigeria. The table indicated that 81.8% of the men and 72.5% of the women complained about decline in fish species and populations. This is because in a small-scale fishing community, many households are involved in fishery-related livelihood activities such as fishing, postharvest fish processing, fish trading, making and mending of fishing materials (OECD 2001). Therefore, decline in fish species and populations will affect their means of livelihood. Also 79.2% of the aged men and 42.9% of aged women in the area complained about water logging and 59% men and 40.6% women had their small animals drown in the water. Moreover, 69.7% men and 40.9% lamented and worried about the deterioration of their dwellings because of heavy rainfall and flood. In most rural coastal communities of Nigeria, many dwellings are built on piles and raft foundations. These houses easily deteriorate during harsh weather conditions and thus affect the habitability of the aged in the region.

However, in some instances climate variability has some positive impacts on livelihood of rural coastal communities. For instance, one of the positive effects of flooding in the rural coastal communities is the increase in fish catch, especially

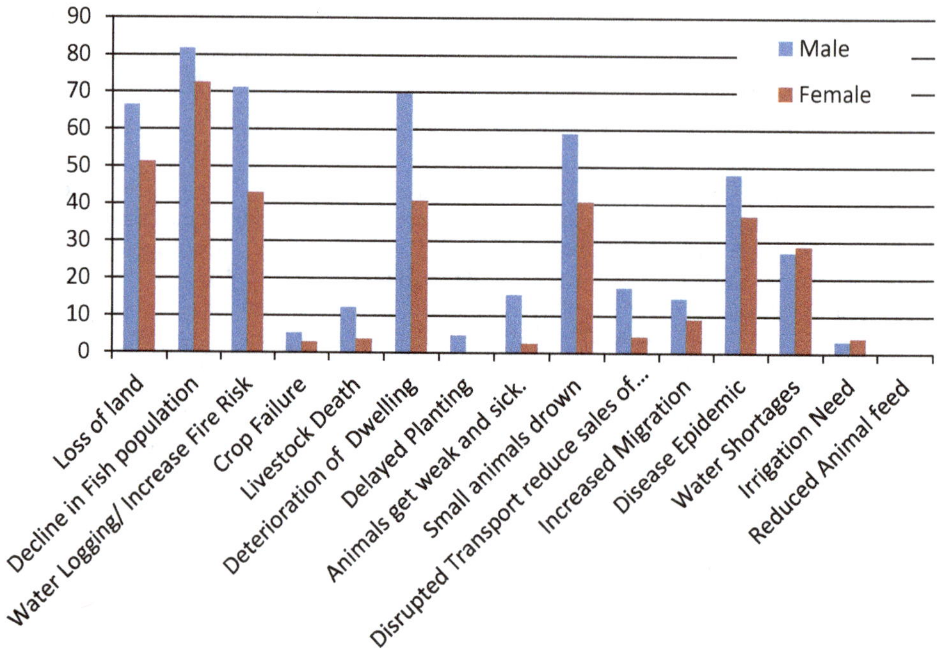

Fig. 2 Climate variability impact of livelihood security of aged men and women in Coastal Zone of Nigeria. (Source: Author's Fieldwork)

during ocean flooding (Fabiyi and Oloukoi 2013). The study revealed that the Atlantic brings large fish to the continental shelf and some large fish are even found in the creeks during extensive flooding. Therefore, climate variability impact is not necessarily negative but can also be positive. Below, are some of the views on the impact of climate variability on their livelihood shared by aged men and women in Ilaje and Eseodo LGAs of Ondo:

> Floodwater does not damage our land and properties in the past, it comes for some few hours and subsides back into the ocean but within the past few years it has been staying longer than a month, destroying lives and properties **(Olayemi Elijah)**
>
> Floods are usually associated with heavy rainfall and ocean tides causing damages to our properties and drowning a number of people and animals. During Heavy rainfall, between June and July houses and land becomes waterlogged. The soil becomes saturated with water making it difficult for us to move from one place to another, disturbs our daily activities and causes retrogression in our socio-economic activities. (**Thomas Ogunmola, Male, 67 years**)
>
> Flood has destroyed our crops, fishes and animals. Many of our animals die because of high salt level in the water. The flood water after mixing with sea water becomes polluted causing purging in human. The crude oil clogs the mat making plant (ifere), degrades the bamboo plantation (Opee), and also kills the aquatic animal, which serves as the major source of income to us (**Osamoluwa Male, 71 years**)
>
> Socio-economic activities comes to a standstill during the flooding period, we women will not be able go to the market to trade. There is also high mortality rate in infants and the pregnant women during child labor because we lack health facilities. The only health centre

available is too far from the community and there is no means of transportation most especially during the time of flooding (**Adeola Ebijiimi Female 75 years**)

In this year, from January to April, the rainfall was so scanty, resulting into heat wave, the weather became unbearable hot and the soil became too dry and too hot to grow crops on, any crop planted at this period dies off, but from May to July the rainfall became so intense which eventually resulted into flooding (**Ebipatei Jonathan**)

We do encounter a lot of health and environmental hazard, these includes: rheumatism, purging, death of animals, degraded water and soil, erosion, drowning of our children and loss of life in general. (**Obele**)

Many of us aged suffer from rheumatism due to extreme cold weather while some of us have lost their lives due to drowning during flooding or tidal surge (**Eyinmiieka Ayebogbon**)

Figure 3 also indicated the impact of climate variability on livelihood security of the rural aged men and women in the Sudan Savannah Zone of Nigeria. The table revealed that 77.2% of the men and 89.3% of the women complained about loss of land. 85.3% of the men and 78.0% of the women complained of reduced pasture for their animals. 78.3% of the men and 71.4% of the women are worried because of water shortage and 67.7% men and 48.9% women felt the need for irrigation. Also, 90.6% of the men and 82.1% of the women lamented about crop failure and 77.2% men and 42.9% lamented about their dying livestock. From all these analysis, it is obvious that livelihood in Sudan Savannah Zone are affected by climate variability impact of drought and irregular rainfall. This is in line with Challinor et al. (2007) who explained that agricultural potential of the Sudan Savanna in West Africa is limited by insufficient water availability due to high rainfall variability and frequent droughts.

Our major occupation here is faming and rearing of livestock animals. Everything is okay. The only change is the issue of rain and drought. Before, rain starts around April and

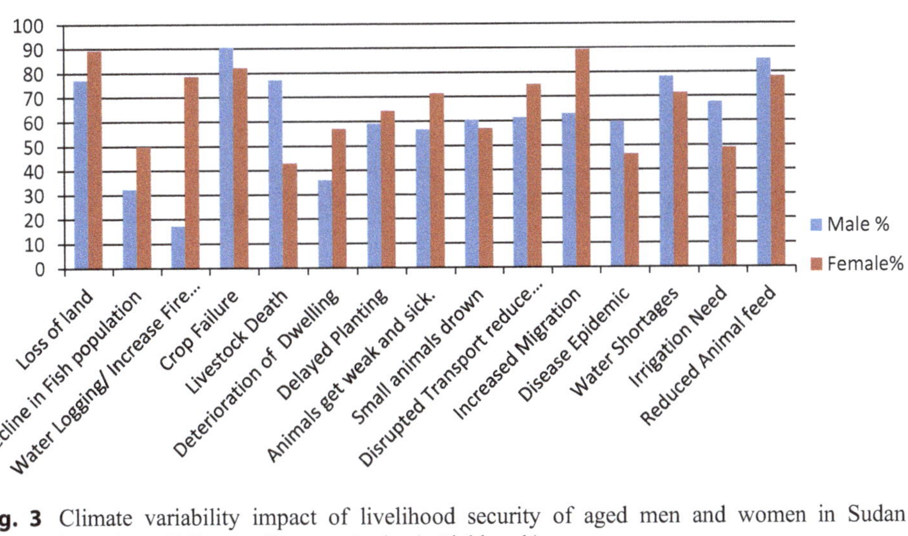

Fig. 3 Climate variability impact of livelihood security of aged men and women in Sudan Savannah Region of Nigeria. (Source: Author's Fieldwork)

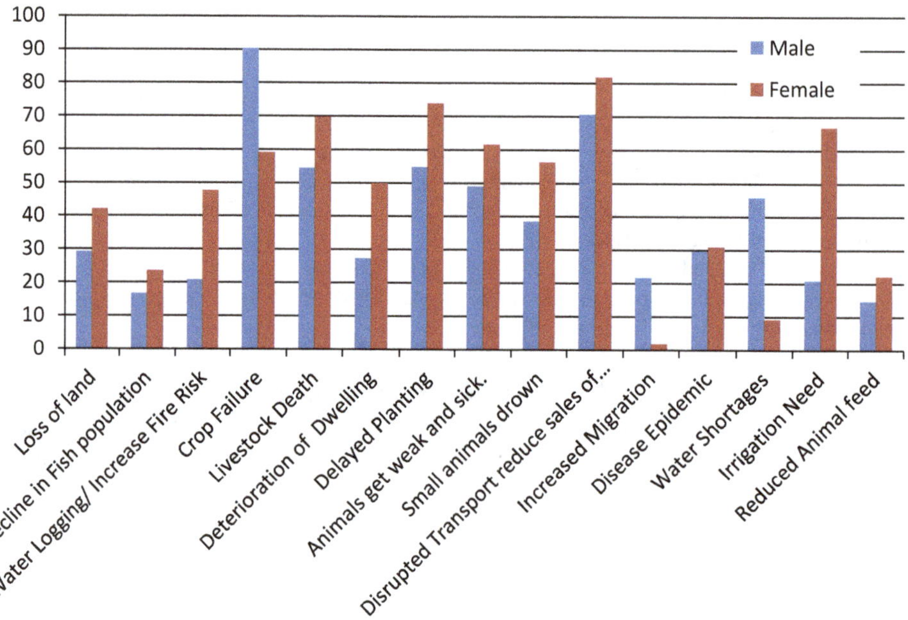

Fig. 4 Climate variability impact of livelihood security of aged men and women in Montane Region of Nigeria. (Source: Author's Fieldwork)

sometimes March. But now, that the climate has changed, it may start around June or July (**Al Haruna**)

We have a river here. Sometimes this river overflows and leads to flood. There is nothing we can do about such a situation because we do not have a choice. (**Aisha Muhammad**)

If you observe the route we followed here, the erosion has affected the bridge there. We have got about 30 bags of cement to do the bridge again. Even as we speak, if rain shouldfall, no one will be able to leave this town until the water level reduces (**Kulu Galadima**)

The drought usually affects our cultivated crops, in most cases we do re-planting after the drought might have destroyed the cultivation. So we plant immediately after the first rainfall that comes after a drought. For the flood, there is nothing we can do about it. What we do in most cases is to gather the redeemable crops and plant again after flood. (**Amina Hakimi**)

Figure 4 reveals the impact of climate variability on livelihood security of the rural aged men and women in the Montane Zone of Nigeria. The table revealed that 70.6% of men and 81.8% of women were not happy about disrupted transport reducing sales of goods; 58% of the men and 59% of the women complained of crop failure; 54.8% men and 73.8% talked about delayed planting; 49% of men and 61.5% of the women worried about their animals getting weak and sick while 54.4% men and 69.8% women complained about their dead livestock. This analysis was further corroborated with comments on the experiences of the aged men and women's climate variability impact on their livelihood security in the different selected villages of Plateau State of Montane Zone of Nigeria.

In this area we do not experience flood, but over there (neighboring village), all their houses and roads are filled with flood (**Sunday Gaarus**)

> Lot of houses are usually affected. But for crops, only rice will do well in such condition. Last year we harvested plenty rice because rice needs a lot of water to grow. But crops like corn, guinea corn did not survive the weather. Millet also, for those farmers that had enough rain harvested it as well. We got millet here last year because we cultivated it around July ending to October ending, because it also can survive heavy rain. But crops like the ones I mentioned earlier, corn, and guinea corn will not survive it (**Zababiya**)
>
> In other places, they have not even had any rainfall at all this year. But we here are very lucky to have experienced several rainfall, though heavy. If we now start complaining about heavy rain, what should other places that have not experienced rain say? We do not know why God has decided to do it this way. So, to tell you I have an idea is unexplainable. Everything is God's doing (**Mrs Ngyuk**)

Generally, the impact and the experience of climate change on livelihood security of the aged men and women is peculiar to each ecological zone. This is because each ecological zone has its own peculiar characteristics. This is in line with Callo-Concha et al. (2013) that says vulnerability to climate change impact is highly correlated with its geographical peculiarities. Therefore, the impact on the respondents is relative to the geographical characteristics.

Conclusion

Climate change is rapidly emerging as one of the most serious global problems affecting many sectors in the world especially rural livelihoods in agriculture, cattle rearing, crop farming, fishing, hunting, mining, tourism, and craft and weaving. It was found that the rural aged populations in Nigeria are highly vulnerable to climate change impact, especially in their livelihood security which was discovered to differ by region and by activity. This is in line with Garai (2014) which assessed climate change impact on the livelihoods of coastal people in the south-western part of Bangladesh revealed the elderly as part of the most at risk and vulnerable to climate change impact because of their inability to adapt to unfavorable environmental condition.

While some areas are generally more sensitive to climate change (Saarinen et al. 2012), some livelihoods and activities are more ecologically dependent and exposed to climate change impact. In consequence, the rural aged experience climate variability impact on their livelihood differently in different ecological zones, and their experience is based on the prevailing climate variability characteristics in each ecological zone.

In the Guinea and Sudan savannah zones, the prevailing climatic variability characteristics are drought, desertification, excessive heat and increased temperature, decline in rainfall, delayed onset of rainfall to mention a few. This is also in line with Ayanlade and Ojebisi (2019) which assessed herders' responses to climate variability and changes in the Guinea Savannah region of Nigeria. The study confirmed climate variability characteristics like excessive heat, reduced and delay onset of rainfall, prolonged dry spell, drought to mention a few in the area. Also in the Coastal zone the prevailing climatic variability characteristics are flooding, earlier onset of rain, erosion, increased frequency in storms. The impact on their livelihood in these zones is mainly associated with livestock death, lack of pastures for herds, scarcity of water, pest invasion, delayed planting crop failure, need for irrigation, water logging, drowning of small animals, human and animal illness. This implies that government planning and decision making related to climate change should take the views of the aged populations into consideration, especially for those rural areas affected most by climate change impact.

Acknowledgments This research was supported by funding from Department for International Development (DfID), in the UK, under the program Climate Impacts Research Capacity and Leadership Enhancement (CIRCLE), implemented by the African Academy of Sciences and the Association of Commonwealth Universities.

References

Adger WN (2003) Social capital, collective action, and adaptation to climate change. Econ Geogr 79(4):387–404

Adger WN, Dessai S, Goulden M et al (2009) Are there social limits to adaptation to climate change? Clim Chang 93(3–4):335–354

Anele D (2012) A brief note on the condition of rural areas in0020Nigeria on January 29, 2012. The Vanguard Newspaper. https://www.vanguardngr.com/2012/01/a-brief-note-on-the-condition-of-rural-areas-in-nigeria/

Aniah P, Kaunza-Nu-Dem MK, Abindaw BA, Millar D (2016) Characterizing and explaining smallholder households views and understanding of climate change in the Bongo District of Ghana. Earth Sci 5(2):26–38

Ashwill M, Flora C, Flora J (2011) Building community resilience to climate change: testing the adaptation coalition framework in Latin America. The World Bank, Washington, DC

Ayanlade A, Ojebisi SM (2019) Climate change impacts on cattle production: analysis of cattle herders' climate variability/change adaptation strategies in Nigeria. Change Adaptat Socio Ecol Syst 5(1):12–23

Callo-Concha D, Gaiser T, Webber H, Tischbein B, Müller M, Ewert F (2013) Farming in the West African Sudan Savanna: insights in the context of climate change. Afr J Agric Res 8(38):4693–4705

Central Bank of Nigeria (2006) Statistical bulletin

Challinor A, Wheeler T, Garforth C, Crauford P, Kassam A (2007) Assessing the vulnerability of food crop systems in Africa to climate change. Clim Change 83:381–399

Collier P, Conway G, Venables T (2008) Climate change and Africa. Oxf Rev Econ Policy 24 (2):337–353

Deschenes O, Greenstone M (2012) The economic impacts of climate change: evidence from agricultural profits and random fluctuations of weather. Am Econ Rev 102(7):3761–3773

Douglas AC, Mills JE, Niang M, Stepchenkova S, Byun S, Ruffinia C et al (2008) Internet addiction: meta-synthesis of qualitative research for the decade 1996–2006. Comput Hum Behav 24:3027–3044

Eboh E (2009) Implications of climate change for economic growth and sustainable development in Nigeria Enugu forum policy paper 10. African Institute for Applied Economics, Enugu

Egbe C, Yaro MA, Okon AE, Bisong FE (2014) Rural people's perception to climate variability/change in Cross River State Nigeria. J Sustain Dev 7(2):25

Fabiyi O, Oloukoi J (2013) Indigenous knowledge system and local adaptation strategies to flooding in coastal rural communities of Nigeria. J Indig Soc Dev 2(1):1–19

Garai J (2014) The impact of climate change on the livelihoods of coastal people in Bangladesh: a sociological study. In: Leal FW, Alves F, Caeiro S, Azeiteiro UM (eds) International perspectives on climate change: Latin America and beyond. Springer, Cham, pp 151–163

Government of Kenya (GoK) (2010) National climate change response strategy. Government Printer, Nairobi
Higgins PAT (2014) Climate change as a public and policy issue. Bull Am Meteor Soc 95:780–781
IPCC (2007) Impact, adaptation and vulnerability: contribution on working group 1 of the Intergovernmental Panel on Climate Change to the third Assessment Report of IPCC. Cambridge University Press, London
Keay RWJ (1949) An example of Sudan zone vegetation in Nigeria. J Ecol 37:335–364
Kebede A, Hasen A, Negatu W (2011) A comparative analysis of vulnerability of pastoralists and agro-pastoralists to climate change: a case study in Yabello Woreda of Oromia Region, Ethiopia. Ethiop J Dev Res 33(1):61–95
Kueppers K (1998) Evaluation of the ligneous strata of the vegetation of the Zamfara reserve. I. In: Prospects of pastoralism in West Africa. Reihe I, vol 25. Tropeninstitut, Giessen, pp 41–47
McLeman R, Hunter LM (2010) Migration in the context of vulnerability and adaptation to climate change: insights from analogues. Wiley Interdiscip Rev Clim Chang 1(3):450–461
Mendelsohn R (2008) The impact of climate change on agriculture in developing countries. J Nat Resour Pol Res 1(1):5–19
O'Brien KL, Eriksen S, Nygaard LP, Schjolden A (2007) Why different interpretations of vulnerability matter in climate change discourses. Clim Pol 7(1):73–88
OECD (2001) OECD Annual Report
Ofoegbu C, Chirwa P, Francis J, Babalola F (2017) Assessing vulnerability of rural communities to climate change: a review of implications for forest-based livelihoods in South Africa. Int J Clim Change Strateg Manag 9(3):374–386
Okonkwo C, Demoz B, Sakai R, Ichoku C, Anarado C, Adegoke J, Amadou A, Abdullah SI (2015) Combined effect of El Niño southern oscillation and Atlantic multi decadal oscillation on Lake Chad level variability. Cogent Geosci 1:1117829
Popoola K, Ajayi S (2019) Rural Ageds' awareness and perception of climate variability in Kanke and Riyom local government areas of Plateau State, Nigeria. Arch Current Res Int 16(1):1–13. https://doi.org/10.9734/ACRI/2019/46174
Saarinen J, Hambira WL, Atlhopheng J, Manwa H (2012) Tourism industry reaction to climate change in Kgalagadi South District, Botswana. Dev South Afr 29(2):273–285
Sarr B (2012) Present and future climate change in the semi-arid region of West Africa: a crucial input for practical adaptation in agriculture. Atmos Sci Lett 13(2):108–112
Scholze M, Knorr W, Arnel NW, Prentice IC (2006) A climate-change risk analysis for world ecosystems. Proc Natl Acad Sci USA 103(35):116–120
Thornton PK, Ericksen PJ, Herrero M, Challinor AJ (2014) Climate variability and vulnerability to climate change: a review. Glob Chang Biol
Ubisi NR, Mafongoya PL, Kolanisi U, Jiri O (2017) Smallholder farmer's perceived effects of climate change on crop production and household livelihoods in rural Limpopo province, South Africa. Change Adapt Socioecol Syst 3:27–38
World Bank (2008) Agriculture for development. World development report. The World Bank, Washington, DC
World Bank Group (2017) The world bank annual report
Zeleke MT, Aberra Y (2014) Determinants of the adoption of land management strategies against climate change in Northwest Ethiopia. ERJSSH 1(1):93–118

Permissions

All chapters in this book were first published by Springer; hereby published with permission under the Creative Commons Attribution License or equivalent. Every chapter published in this book has been scrutinized by our experts. Their significance has been extensively debated. The topics covered herein carry significant findings which will fuel the growth of the discipline. They may even be implemented as practical applications or may be referred to as a beginning point for another development.

The contributors of this book come from diverse backgrounds, making this book a truly international effort. This book will bring forth new frontiers with its revolutionizing research information and detailed analysis of the nascent developments around the world.

We would like to thank all the contributing authors for lending their expertise to make the book truly unique. They have played a crucial role in the development of this book. Without their invaluable contributions this book wouldn't have been possible. They have made vital efforts to compile up to date information on the varied aspects of this subject to make this book a valuable addition to the collection of many professionals and students.

This book was conceptualized with the vision of imparting up-to-date information and advanced data in this field. To ensure the same, a matchless editorial board was set up. Every individual on the board went through rigorous rounds of assessment to prove their worth. After which they invested a large part of their time researching and compiling the most relevant data for our readers.

The editorial board has been involved in producing this book since its inception. They have spent rigorous hours researching and exploring the diverse topics which have resulted in the successful publishing of this book. They have passed on their knowledge of decades through this book. To expedite this challenging task, the publisher supported the team at every step. A small team of assistant editors was also appointed to further simplify the editing procedure and attain best results for the readers.

Apart from the editorial board, the designing team has also invested a significant amount of their time in understanding the subject and creating the most relevant covers. They scrutinized every image to scout for the most suitable representation of the subject and create an appropriate cover for the book.

The publishing team has been an ardent support to the editorial, designing and production team. Their endless efforts to recruit the best for this project, has resulted in the accomplishment of this book. They are a veteran in the field of academics and their pool of knowledge is as vast as their experience in printing. Their expertise and guidance has proved useful at every step. Their uncompromising quality standards have made this book an exceptional effort. Their encouragement from time to time has been an inspiration for everyone.

The publisher and the editorial board hope that this book will prove to be a valuable piece of knowledge for researchers, students, practitioners and scholars across the globe.

List of Contributors

Taiwo B. Ayinde
Samaru College of Agriculture, Division of Agricultural Colleges, Ahmadu Bello University (ABU), Zaria, Nigeria

Benjamin Ahmed
Department of Agricultural Economics, ABU, Zaria, Nigeria

Charles F. Nicholson
Nijmegen School of Management, Radboud University, Nijmegen, Netherlands

Martin Munashe Chari
Department of Geography and Environmental Science, University of Fort Hare, Alice, South Africa
Risk and Vulnerability Science Centre (RVSC), University of Fort Hare, Alice, South Africa

Hamisai Hamandawana
Department of Geographical Information Systems (GIS) and Remote Sensing, University of Fort Hare, Alice, South Africa

Leocadia Zhou
Risk and Vulnerability Science Centre (RVSC), University of Fort Hare, Alice, South Africa

Mohamed Meddi
Ecole Nationale Supérieure d'Hydraulique de BLIDA, Boufarik, Algeria

Saeid Eslamian
Department of Water Engineering, College of Agriculture, Center of Excellence on Risk Management and Natural Hazards, Isfahan University of Technology, Isfahan, Iran

Julie Greenwalt
Go Green for Climate, Amsterdam, Netherlands

Michael Dede, Ibinabo Johnson, Prince Nosa and Abi Precious
Chicoco Collective, Port Harcourt, Nigeria

Barbara Summers
CMAP, Port Harcourt, Nigeria

C. M. Githunguri
Kenya Agricultural and Livestock Research Organization (KALRO) Food Crops Research Centre Kabete, Nairobi, Kenya

E. N. Njiru
KALRO Katumani, Machakos, Kenya

Lilian Motaroki
International Centre of Insect Physiology and Ecology (ICIPE), Nairobi, Kenya

Gilbert Ouma and Dorcas Kalele
Institute for Climate Change and Adaptation (ICCA), University of Nairobi, Nairobi, Kenya

Joseph P. Gweyi-Onyango
Department of Agricultural Science and Technology, Kenyatta University, Nairobi, Kenya

Michael Ajanja Sakha and Joyce Jefwa
Botany Department, National Museums of Kenya, Nairobi, Kenya

Josephine Mbandi
School of Computational and Communication Sciences and Engineering (CoCSE), Nelson Mandela Institution of Science and Technology (NM-AIST), Tengeru, Arusha, Tanzania

Michael Kisangari
Centre of Excellence in Information Technology (CENIT@EA), Arusha, Tanzania

K. Sian Davies-Vollum, Debadayita Raha and Daniel Koomson
Environmental Sustainability Research Centre, University of Derby, Derby, UK

Kehinde Olayinka Popoola
Department of Urban and Regional Planning, Obafemi Awolowo University, Ile-Ife, Nigeria

Anne Jerneck
Lund University Centre for Sustainability Studies, Lund, Sweden

Sunday Adesola Ajayi
Department of Crop Production and Protection, Obafemi Awolowo University, Ile-Ife, Nigeria
Institute for Sustainable Development, First Technical University, Ibadan, Nigeria

Index